EDIBLE MUSHROOMS

Barbro Forsberg   Stefan Lindberg

EDIBLE MUSHROOMS

Safe to Pick, Good to Eat

FACTS CHECKED BY MICHAEL KRIKOREV

TRANSLATED BY ELLEN HEDSTRÖM

SKYHORSE PUBLISHING

Library of Congress
Cataloging-in-Publication Data
Forsberg, Barbro, 1950-
  [Matsvampar. English]
  Edible mushrooms : safe to
pick, good to eat / Barbro
Forsberg ; photographs
by Stefan Lindberg ; facts
checked by Michael Krikorev ;
translated by Ellen Hedstrom.
    pages cm
  ISBN 978-1-62873-644-1
(paperback)
  1. Edible mushrooms--
Identification. 2. Mushrooms,
Poisonous--Identification. I.
Title.
  QK617.F5713 2014
  579.6'1632--dc23

2013045148

SHAGGY INK CAP.
PAGE 2, PARASOL MUSHROOM.

English Translation © 2014
by Skyhorse Publishing

First published in 2012 as
Matsvampar by Barbro Fors-
berg and Stefan
Lindberg, Bonnier Fakta, Sweden

Text, graphic design, and scanning of
mushrooms by Barbro Forsberg

Photography and reprographics by Stefan Lindberg

Facts checked by Michael Krikorev

Skyhorse Publishing books may be
purchased in bulk at special discounts
for sales promotion, corporate
gifts, fund-raising, or educational
purposes. Special editions can also
be created to specifications. For
details, contact the Special Sales
Department, Skyhorse Publishing, 307 West 36th
Street, 11th Floor, New York, NY 10018 or info@
skyhorsepublishing.com.

Skyhorse® and Skyhorse Publishing® are registered
trademarks of Skyhorse Publishing, Inc.®, a Delaware
corporation.

www.skyhorsepublishing.com

10 9
ISBN: 978-1-62873-644-1
Printed in China

Images on cover are not to scale.

Contents

A LOVE OF PICKING.............1

MUSHROOMS FOR PICKING...2

MUSHROOMS IN THE
    KITCHEN.....................4

MUSHROOM TERMINOLOGY...6

Porcini..........................9

Summer Cep.................17

Pine Bolete...................25

Bay Bolete....................29

Slippery Jack..................33

Weeping Bolete..............39

Velvet Bolete.................43

Orange Birch Bolete.........49

Red-capped Scaber
    Stalk........................53

Birch Bolete..................57

Chanterelle....................61

Cantharellus pallens............69

Trumpet Chanterelle.........73

Yellow Foot...................81

Black Trumpet...............87

Wood Hedgehog.............95

Terracotta Hedgehog........99

Sheep Polypore.............103

Giant Puffball...............107

Warted Puffball.............111

Pig's Ear.........................115

Clustered Coral..............119

Wood Cauliflower..........123

Black Morel...................131

Weeping Milk Cap..........135

Saffron Milk Cap...........143

False Saffron Milk Cap....147

Bare-toothed Russula.......151

Copper Brittlegill...........155

Yellow Swamp Brittlegill...159

Russula romellii................163

St. George's Mushroom....167

Arched Wood Wax.........173

Herald of Winter............177

Fairy Ring Mushroom......181

Gypsy..........................187

Slimy Spike Cap..............191

Parasol Mushroom.........195

Shaggy Ink Cap.............201

Burgundy Truffle...........205

POISONOUS MUSHROOMS..208

ACKNOWLEDGMENTS......214

HELPFUL BOOKS FOR
    MUSHROOM
    IDENTIFICATION..........214

ALPHABETICAL INDEX.......216

A LOVE OF PICKING

THE HUMIDITY IS *high and our pots and pans simmer away. Scores of mushrooms are cleaned and left to soak while our woodstove is working at maximum capacity.*

Suddenly, a dragging sound from the larder makes us raise our heads from the mounds of mushrooms.

Swish, swoosh . . . splat!

What are these strange, poltergeist sounds? They have a rational explanation; steam from the boiling mushrooms causes the wallpaper to release rapidly from the stone wall. One strip has already piled on the floor, and several more have come loose up near the ceiling. They look like they're about to fall down at any minute. Soon the whole wall is bare; undressed, hot, and all shiny from the damp.

THIS IS JUST one of many mushroom-collecting memories. Simply put, we are mushroom maniacs and we love picking, cleaning, and eating mushrooms.

It's exciting to test edible mushrooms that you've never tried before, but sometimes it can be hard to identify them using regular mushroom guides. This is why we've made a book that's slightly different. We've included forty edible mushrooms that we have gotten to know over the years, and we hope that the average mushroom picker will dare to try some of these mushrooms that they might not have tasted before. These fungal treasures create such heavenly tastes.

Barbro Forsberg
Stefan Lindberg

MUSHROOMS FOR PICKING

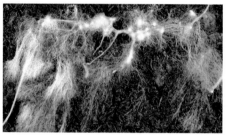

A Velvet Bolete's fruiting body.

Under the earth's surface we find the vegetative part that creates the fruiting bodies that are picked. The white stuff is the mycelium of a Velvet Bolete that is growing out from the roots of a pine tree, giving the tree an increased rate of nutrient absorption.

If you are unsure of the type of mushroom you have, you can determine the color of the spores. Place the mushroom with the surface that creates the spores face down on black or white paper (depending on the color you are expecting the spores to be). Cover the mushroom so the spores don't fly around the room. After a few hours, a deposit of spores will have fallen onto the paper. The photo to the left shows the spores from the St. George's mushroom, which are white, in contrast to its poisonous look-alike, the Livid Agaric, which has a pink spore print.

IN THIS BOOK we have selected forty mushrooms that are safe to pick and that taste good. That's not very many when you consider that there are 5,000 mushroom types with a visible fruiting body (the part of the mushroom that is picked) in the United States alone and of these, only a few hundred are considered to be good, edible mushrooms.

All mushrooms consist of a network of fine threads underneath the earth, the mycelium, which obtain nutrients via organic material, dead or alive.

Mushrooms are divided into different groups depending on where they get their nutrients. Mushrooms that live off of, and break down, dead organic matter are called saprophytes and they play a vital part in nature. Within this group there are several tasty edible mushrooms, such as the St. George's and the Parasol Mushroom.

Mycorrhizal mushrooms have a symbiotic relationship with living plants, which are their hosts. The mycorrhizal mushrooms get their food from these plants in exchange for providing the plant with water and minerals. The mushroom's mycelium works as an extension of the plant's root system, thereby creating a successful symbiosis between plant and mushroom.

Most edible mushrooms are mycorrhizal mushrooms and are attached to various types of trees. The Chanterelle, for example, grows with both coniferous and deciduous trees while the Slippery Jack only has one host, the pine tree.

The third group, the parasites, include only a few edible mushrooms. This book only contains one, the Wood Cauliflower on page 129.

A lone mycelium can produce a vast number of fruiting bodies within a limited area and it is here that the spores are created and dispersed to create new mushrooms.

Determining the Species

THE SPECIES OF mushrooms that are unique in their appearance are easy to learn, while those that have several look-alikes take a bit longer to get to know.

It's a good idea to own several mushroom books so you can compare the pictures and descriptions. Make sure the books are less than ten years old, as new discoveries are constantly being made and some mushrooms might be re-evaluated and defined as edible where they were once inedible and vice versa.

If you use all your senses, i.e. memorize the smell and the feel of each mushroom, you will eventually get to know a group of edible mushrooms and you can skip poring through your guidebook every time you come home from picking mushrooms.

A spore print is always a good idea if you want to be completely sure that it's not a poisonous look-alike and you can find out how to do this in the photo and text at the bottom of page 2.

It's harder to distinguish between some types of gilled mushrooms (agaricales), especially *Russulas*. Many look so alike that it can require microscopic examination to be completely sure. Thankfully, there are

few poisonous *Russulas* but some of them taste bad and are not considered edible. How to recognize and taste test a *Russula* is described on page 159.

We have chosen not to present the edible wild white meadow mushrooms (*Agaricus*) in this book (and yes, there are poisonous white mushrooms). This is mainly due to the fact that each year there are several mistakes made between the Destroying Angel and wild white meadow mushrooms. Besides, you can just as easily buy the cultivated ones cheaply in the supermarket.

According to the food advisory board, neither wild nor cultivated white mushrooms should be over-consumed.

Locating Mushrooms

IF YOU CAN learn which host tree goes with which mushroom, it will be easier to find your favorite mushrooms.

Many mushroom species grow over a long fruiting period and can even appear several times in a season, making it worth returning to the patch where you found them.

You stand a greater chance of finding mushrooms in an older forest where the ground has not been disturbed by tree felling for the past forty or fifty years. Here, you'll often find a wider variety of mushrooms. Some areas of nature are more unique and are host to rare animals, plants, and mushrooms. For example the Pig's Ear on page 115 is a so-called "indicator mushroom," which indicates that the forest has not been recently disturbed.

Another tip is to join an arranged mushroom tour with a knowledgeable guide and in this way learn more about how to locate and recognize mushrooms.

MUSHROOMS IN THE KITCHEN

AS SOON AS you've plucked the mushroom from the ground, you need to treat it like fresh produce and take care of it as quickly as possible. You should always do the first, rough cleaning as you pick it, removing the soil-covered base, and if you are able to finely clean it the same day, it will last longer. More information on picking and cleaning can be found on page 13.

If you don't have time to prepare the mushroom after finely cleaning it, it will keep for a few days in the refrigerator. If you have a large quantity of mushrooms and the weather is cool, you can store the cleaned mushrooms outside under a roof for 24 hours.

Preparing Fresh Mushrooms

BEFORE YOU PREPARE the mushrooms, you need to remove excess liquid. After cleaning them, place in a frying pan or pot with no cooking fat or oil, at a low heat. If the mushrooms are dry, you can add a dash of water. Once the mushrooms start to release liquid, increase the heat; allow the liquid to evaporate before adding butter. You can then salt and fry the mushrooms. After this, the mushrooms can be cooked further, for example in a stew, sauce, or a soup.

Parboiling and Freezing

WHEN FREEZING MUSHROOMS you don't need to remove all the liquid as you would when preparing fresh mushrooms, and no butter needs to be used. When frozen, the mushrooms should be covered in their own juices, as this makes them less chewy. They will keep for up to a year.

You can also freeze the mushrooms directly without parboiling them. They take up more space but will taste fresher because you can cook them without having to first defrost them. Place the frozen mushrooms straight into a hot frying pan with some butter and fry quickly.

Drying

THE BEST WAY to store mushrooms is to dry them, as the flavors will be concentrated and they will last pretty much forever if stored in a sealed glass jar in a dark, dry place.

You can dry mushrooms in several ways. The main thing is that they need to air out, and a kitchen table makes a good surface for this. Cover it with newspaper to protect the table's surface, and place a clean cloth on top, like an old table cloth or a sheet. If you place the mushrooms straight onto the newspapers they can stick. Wooden frames with mosquito netting also work well. The mushrooms should be spaced out slightly.

Mushrooms that are less "meaty" are easiest to dry. More dense mushrooms such as the Porcini should be cut into thin slices before drying.

Some mushrooms are considered less suitable for drying—Chanterelles and Hedgehog mushrooms, for example—as they become chewy and bitter, so parboil and freeze these instead.

Drying time is normally two to four days at normal room temperature. When they're ready, the mushrooms should be crispy and easy to break or crumble. You can even buy electric mushroom dryers or electric dehydrators that will dry the mushrooms in a few hours.

Drying them in the oven is not a good method; in order to get the liquid out, you would need to leave the door open, which makes the heat uneven. At temperatures above 104°F (40°C) the mushrooms will start to cook and will be destroyed.

If you have a lot of dried mushrooms, you can make mushroom flour using a food processor or a mortar and pestle. The flour tastes great when added to sauces and soups.

Don't forget that mushrooms release spores while drying, and in larger quantities can cause problems in small children or those with allergies or asthma.

Soaking

BEFORE COOKING WITH dried mushrooms, you need to soak them. Boil some water and pour into a bowl. Dilute with some cold water until it is lukewarm and not above 104°F (40°C). Make sure your fingers are dry when removing the mushrooms from the jar and don't stand too close to the steam from the pot.

The mushrooms should be just covered by water and after roughly 15–30 minutes they are ready to use.

You can also use the soaking liquid to add some flavor to your dish. Add a little at a time and taste it as you go; otherwise it can become too bitter.

Two pounds of mushrooms (1 kg) is equivalent to approximately 3½ oz (100 g) of dried mushrooms.

A mosquito net is ideal for drying mushrooms. Here, a few thinly sliced Porcinis are lying out to dry.

Dried mushrooms need to be stored in a dark, dry place. New research shows that edible mushrooms are a lot healthier than once thought. Among other things, they contain antioxidants and minerals.

MUSHROOM TERMINOLOGY

VARIOUS MUSHROOMS HAVE the most amazing shapes and colors and even one species will often go through a major transformation during its lifespan.

To simplify things when defining the mushroom species there are certain terms for the different appearances—mainly concerning the shape of the caps.

The spore-producing tissue, the hymenium, is also an important distinguishing feature. This is usually found under the cap on the pores, ridges, teeth, and gills, and the shapes all have different names.

Spore-producing Tissue

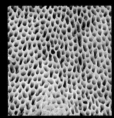
Tubes/pores
(Slippery Jack)

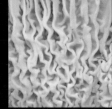
Ridges (Chanterelle)

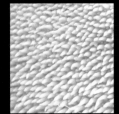
Teeth
(Wood Hedgehog)

Gills/Lamellae
(Parasol Mushroom)

Cap Shapes

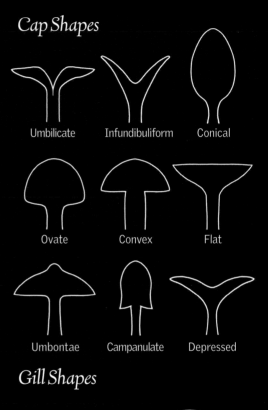

Umbilicate　　Infundibuliform　　Conical

Ovate　　Convex　　Flat

Umbontae　　Campanulate　　Depressed

Gill Shapes

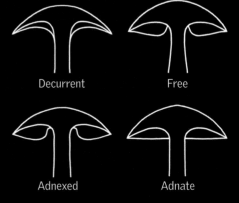

Decurrent　　Free

Adnexed　　Adnate

EACH CHAPTER in this book ends with a page presenting a short overview of the species. The following is a general explanation of the descriptions.

FRUITING BODY: The part of the mushroom that grows above the earth (except truffles).

CAP: The top part of the mushroom.

GILLS, RIDGES, PORES, AND TEETH: The hymenium, or surface where the spores are produced, is usually found underneath the cap.

STIPE: The stem of the mushroom, usually upright, occasionally angled.

FLESH: The inside of a mushroom.

VEIL, RING, VOLVA: The veil is a thin layer that protects the young fruiting body and breaks as the mushroom grows. The partial veil, attached to the center of the stipe and the edge (margin) of the cap protects the young gills or pores. The remnants create a ring around the stem of the mushroom. The universal veil envelops the whole fruiting body and the remnants create a "cup" at the base of the stipe called a volva, and sometimes scales on top of the cap.

SPORE PRINT: The color of the spores is a good way to identify mushrooms. The spore powder can be seen on gills and at the pore openings. Determining the color of the spores can help to identify the type of mushroom it is (page 2).

RANGE AND HABITAT: Different kinds of mushrooms appear in different seasons, types of terrain, host trees, and weather conditions. Some species are more particular than others and are therefore sometimes rarer. These should be spared from picking.

SMELL AND TASTE: It is very useful to be able to recognize the smell of a mushroom. Smell the spore-producing tissue, the hymenium, as this contains most of the aroma producing molecules. We do not recommend tasting raw mushrooms with the exception of the *Russula* (page 153).

PICKING AND CLEANING: It is best to place picked mushrooms in an airy basket with a clean piece of newspaper at the bottom. Do the dirty work on site with a knife and brush. Cut off the earth-covered stipe and, if necessary, divide the mushroom before it is placed in the basket. Avoid placing heavy fruiting bodies on top of fragile mushrooms. Never use plastic bags, as this will increase the rate of bacterial growth. Unknown mushrooms should be kept separately and if in doubt, do not eat them.

PREPARING AND STORING: Most edible mushrooms are suitable to fry and stew. If you wish to preserve the mushroom you should dry or parboil them before freezing (page 3).

LOOK-ALIKE MUSHROOMS: Most mushrooms have a look-alike and most of these are not harmful. Some are just as good as edible mushrooms and some aren't. A few look-alike mushrooms are mildly to deadly poisonous (page 208).

PORCINI

◁ Soon we will be enjoying a Porcini-filled sandwich, a delicacy that's hard to beat.

▢ Just like freshly baked bread, a perfectly "baked" Porcini cap.

THE WINDOWS HAVE been rolled down, and as the car slowly rolls down the country road, we carefully scan the surrounding area, peering through the trees.

"Stop!" I cry, "I think I see some mushrooms!"

We push the thickly-growing spruce branches aside and stumble into a beautiful fairy tale forest where the setting summer sun shines its golden rays. This golden light reveals every mushroom picker's dream and we find ourselves in a treasure trove filled with Porcini, the king of mushrooms.

Large and small caps appear, like smooth, freshly baked buns, and they cluster in the moss on their chubby little rococo stalks.

I PLACE MY fingers around a balloon shaped cap, a Porcini "baby," and feel the moss against my fingers. I clutch the end of the stalk and gently twist to remove the whole fruiting body. It is heavy and dense, despite not being fully grown, and half the stalk is still hidden inside the almost spherical cap. I peel away the earth around the base of the stalk to reveal a flesh devoid of any insect damage—a real find!

A pretty vein-like pattern is faintly visible on the round stalk, especially near the top. The thin pore surface, which is puffy and yellow-green in the older specimens, is crispy and hard and a light beige color. The smell is pleasant, mild and nutty and a little sweet and sour. Even the older Porcinis seem unscathed by attack from maggots and insects.

The cap surfaces are sticky from the rain, which means that these boletes have grown at record speed.

WE NEED TO thoroughly clean and cut away anything that's infested with insect larvae. The larger mushrooms are divided and placed in their own basket so they don't squash the smaller ones. The pores on the older specimens are soft and slimy; they will only turn to mush in the pan, so we peel away the soft tubes and discard them.

The baskets are soon filled with these majestic mushrooms; this is what we call a Porcini year . . . and mushroom happiness!

Some years are Porcini years, and that's when the woods kindly offer up vast amounts of this amazing edible mushroom.

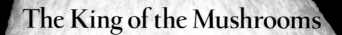

The King of the Mushrooms

THE PORCINI RULES the mushroom kingdom. This is not just because of its size, as the Porcini is the most prized and frequently picked edible mushroom in the world. It was even feted during Roman times. It's also one of the most commercially used of all the wild edible mushrooms.

Millions of tons of Porcini are shipped around the world every year. Freshly frozen, dried, pickled, flour, and so on. China is one of the largest exporters, and it has an extensive trade in growing wild edible mushrooms. Italy and France are some of the largest importers; despite having Porcini mushrooms on their own soil, the natural supply simply is not large enough.

Cep, another name for Porcini, is really an umbrella term for several species, for example the Summer Cep and Pine Bolete. In the United

The Bolete Eater, *Hypomyces chrysospermus*, is a common parasitic mold that is found on ceps. Usually, at first you don't see the attack on the thick base of the stipe. Eventually, the mushroom becomes covered by a white coat that turns yellow and gives off a bad smell.

States, several closely related species are referred to as the King Bolete group. It doesn't really matter if you confuse these species, as they are all edible and each tastes just as good as the others.

Porcini grows mainly in coniferous and deciduous forests in the northern hemisphere and grows well with many different types of trees.

The species can now also be found in South Africa, where it unintentionally travelled with imported trees.

THE NORTHERN EUROPEAN Cep is considered the best tasting cep. It has a sweeter and fuller taste than those from other parts of the world.

In the United States, Porcini grows from mid-summer to late fall all over much of the country, in forested regions and in suburbs with planted trees. Some years the woodlands are teeming with ceps during a limited but intensive growth period—usually one to two weeks. You need to be quick, though, as you are not only competing with insects for these delicacies; even larger animals like deer, wild hogs, squirrels, and mice, are fond of ceps.

In parts of North America, there is a variety known as the King Bolete and even in Sweden the name has royal ties (the Swedish name for it is "Karljohan"). Our first king to carry the name Bernadotte was Karl XIV Johan, who became king of Sweden and Norway in 1818. He brought French culinary traditions to Scandinavia and taught the upper class to eat "tube cep" or "gentleman's mushroom," as it was then called.

It was much later that the general population found an interest in mushrooms as a food source.

The tube cep, also known as the gentleman's cep, is a large mushroom. This species distinguishes itself from other mushrooms in that the hymenophore, on the underside of the cap, does not consist of gills but closely packed barrels or tubes and this entire layer can easily be distinguished from the meaty cap. Because of its bountiful existence and its meaty consistency, this is one of the most treasured mushrooms.

From The Art of Modern Cooking (Modärn kokkonst), *1909*

Porcini *Boletus edulis*

FRUITING BODY: Short and thick. Can grow quite large and weigh more than 2 lb (1 kg).

CAP: 2–8 inches (5–20 cm) wide. At first convex and then almost completely flat. Sticky when damp. The cap color varies from light to dark brown, finishing off with a white margin.

The cap color can vary from light to dark brown.

FLESH: White. The flesh of the cap starts off firm and then softens. Nearer the skin of the cap, the flesh is always a faint brown/red shade (page 19). The flesh of the stalk is firmer and more fibrous than the cap.

PORES: Start off a light gray/beige and then turn yellow to green-yellow.

STIPE: 2–6 inches (5–15 cm) long. Has a white network pattern on a darker surface that becomes more fine and visible toward the cap. The shape is usually chubby and swollen but can also be slim.

SPORE PRINT: Olive brown.

RANGE AND HABITAT: Grows with trees in coniferous and deciduous woods all over cooler regions of North America during the summer and fall. Some years in vast amounts.

SMELL AND TASTE: Nutty or almond-like. Mild and pleasantly sweet.

PICKING AND CLEANING: Avoid wet weather, as the Porcini soaks up water and is quickly attacked by insects and snails. Roughly clean when you pick them, as you often need to remove quite a lot of the mushroom. Only use the ones that have a pleasant smell. If it has been attacked by the Bolete Eater it will smell bad and is not good for you (page 20).

PREPARING AND STORING: Drying is the best way to conserve ceps. Cut the mushroom as thinly as possible and use an electric dehydrator or place them sparsely on a mosquito net in a warm, dry area. If you have a lot of space, you can freeze fresh mushrooms too. When cooking, cut bits of the frozen mushroom straight into the pan and it will taste as fresh as when it was picked.

LOOK-ALIKE MUSHROOMS: Summer Cep, *Boletus reticulatus* (page 23), Pine Bolete, *Boletus pinophilus* (page 31), and Bay Bolete, *Xerocomus badius* (page 35) are equally tasty edible mushrooms. However, the Bitter Bolete, *Tylopilus felleus*, tastes very bitter (page 29).

SUMMER CEP

The Summer Cep has varying shades depending on the weather and where it grows; from red/brown to a light chamois color.

WE GAZE THROUGH beech tree leaves and lilies of the valley. Every year they arrive—the delicacies that can be found among the beech trees on the slope down to the pond. Sometimes you even find them around midsummer but this dry spring and early summer have delayed the Summer Cep.

Finally, at the beginning of August, the weather has changed and after several thunderstorms and much heavy rain, it's only a matter of time before the mushrooms will appear.

We have been looking for these "champagne corks" for the past few days—this is what the professional chefs call the tiniest examples of these mushrooms, and these are the ones that taste the absolute best.

YES, RIGHT THERE by the tree stump are two, one with a large convex-shaped, light beige cap and a small, chubby one with a darker cap; they weren't here yesterday.

We peel the earth from the base of the stipes and the flesh is crispy and hard; it feels almost like cutting into raw potatoes.

The larger ones have already been attacked by maggots, but the smaller mushrooms are perfect all the way down to the base of the stipe—a perfect champagne cork for our omelet lunch.

Smaller Summer Ceps can be more infested than larger ones.

An Early Delicacy in the Deciduous Forest

The tough flesh on the Summer Cep can get quite damaged and cracked during drought.

THE SUMMER CEP is often confused with the similar Porcini mushroom. It doesn't really matter, as they're both fantastic edible mushrooms.

Summer Cep grows in the Midwest and Northeast United States. It thrives with beech and oak, and can appear early on in summer, sometimes as early as June and July, fruiting into early fall.

The species is one of our tastiest edible mushrooms. In some countries, the Summer Cep is considered even better than the Porcini, which is probably because the flesh has a slightly firmer consistency.

Summer Cep *Boletus reticulatus*

FRUITING BODY: Short and fat. Can become vast and weigh over 2 lb (1 kg), just like the Porcini.

CAP: From light beige to red/brown. Starts off ovate with a curved margin, becoming convex and approximately 4–8 inches (10–20 cm) wide. Has a dry cap surface which is finely downy like chamois. With age, often cracks in dry weather, which exposes the white flesh of the cap.

FLESH: White and firm.

PORES: At first white and crispy, then soften and turn yellow to yellow-green.

STIPE: 3–5 inches (8–12 cm) long, thick and usually in a lighter shade than the cap with a vein-like network that usually goes all the way to the base of the stipe. It can change between white and brown on the same fruiting body and easily cracks in drought.

SPORE PRINT: Olive brown.

RANGE AND HABITAT: Grows mainly during the summer months in the Midwest and Mid-Atlantic States, sometimes appears in early June. Thrives in deciduous forests, especially with oak and beech trees. Several related edible species are widespread.

SMELL AND TASTE: Mildly nutty. When cooked it compares with Porcini.

PICKING, CLEANING, PREPARING, AND STORING: See Porcini on page 21.

LOOK-ALIKE MUSHROOMS: Porcini, *Boletus edulis* (page 15), and Pine Bolete, *Boletus pinophilus* (page 23), are equally tasty edible mushrooms. Bitter Bolete, *Tylopilus felleus*, is an inedible Cep. It tastes very bitter and will ruin the whole dish. The fluffy pore openings are at first light gray, then pink. The stipe is dark and patterned with a lighter belt nearer the cap.

Bitter Bolete grows in conifer forests, usually on decaying wood.

PINE BOLETE

THE CLEARING IS *a few years old, strewn with eagle fern and small, prickly spruce. The terrain is hard to navigate and our destination is the other side where a fantastic old forest awaits us.*

Suddenly, a moose appears from the scrub and blocks the path, but after a few moments of curious staring, he decides to move on.

WE ARE SURROUNDED *by 200–300-year-old pine and spruce trees, some of them 100–130 ft (30–40 m) tall. Not only do you feel humbled when gazing at the tree tops, you also get quite dizzy. But maybe our moose encounter contributed to this dizziness!*

Mushrooms love this type of old, virgin forest and the place is teeming with different species in various shapes and colors, most of which we don't have a clue as to their identity.

Soon a familiar face appears, a large Pine Bolete, and the fruiting body is firm and feels good. We remove the pore surface, and the cap flesh is surprisingly free from infestation. Even the stipe is in good shape.

One single mushroom filled our basket and weighed in at 3 pounds (1.4 kg).

A fully grown Pine Bolete. When you remove the tubes you can see how much of the flesh has been infested.

A Mushroom of Distinction

THE PINE BOLETE has several names in Swedish and is also known as the Pinewood King Bolete in English. The fact that is has "pine" in its name is very fitting, as it usually grows alongside the pine tree. As an edible mushroom, it's on par with the Porcini and the Summer Cep.

This species is found spread out across the northern hemisphere, and it thrives in coniferous forests with lean, sandy soil. In Sweden, it grows only with pine trees. The Pine Bolete's fruiting bodies can sometimes grow very large and in a few, rare cases can weigh several pounds.

Pine Bolete *Boletus pinophilus*

FRUITING BODY: Like several other ceps, it can grow quite large when fully mature and weigh several pounds.

CAP: Approximately 4–8 inches (10–20 cm) wide. Stays convex for a while but eventually flattens out. Often irregular in shape when mature. The cap surface is a deep red/brown or copper color. The surface is hard, dimpled, and rough and sticky when damp and the flesh is faintly red/brown just underneath the cap surface—just like the Porcini (page 19).

PORES: Lightly gray-beige and then deep yellow. Finally, they achieve a red/brown shade at the openings.

STIPE: Approximately 3–5 inches (8–12 cm) in height. Thick and egg or pear shaped when young. The network pattern is dark brown at the bottom and gets lighter toward the cap. The color is almost as deep as the cap's.

FLESH: White and firm, both in the cap and stipe.

SPORE PRINT: Olive brown.

RANGE AND HABITAT: Widely distributed across the United States in association with pine trees.

SMELL AND TASTE: Nutty or almond-like, mild and pleasantly sweet.

PICKING, CLEANING, PREPARING, AND STORING: See Porcini on page 21.

LOOK-ALIKE MUSHROOMS: Summer Cep, *Boletus reticulatus* (page 23), and Porcini, *Boletus edulis* (page 15), are equally tasty edible mushrooms. The Bay Bolete, *Xerocomus badius* (page 31), is also a good quality edible mushroom. Beware the foul tasting Bitter Bolete, *Tylopilus felleus* (page 23).

Porcini is an edible look-alike.

BAY BOLETE

The Bay Bolete is hardy against the cold and appears either alone or in small groups. The pores, which turn a bluish color when touched, do not grow all the way to the stipe.

THE MILD ATLANTIC air and a brave autumn sun give nature some time to breathe after frosty nights, and our mushroom patches tempt us yet again. A few species, apart from the Trumpet Chanterelle, should still be around, even if November is just around the corner.

With a brown, limp cap and puffy pores that hang down, we see a Bay Bolete in solitary glory. Close up, hues of purple can be seen on the cap surface, which is reminiscent of suede, and the cap is wavy and stretched at the margin. The brown, streaky stipe looks too weak to carry the magnificent looking cap.

The fruiting body is a miniature version of the Pine Bolete, but the Bay Bolete is a real fighter that can survive a few nights of below-freezing temperatures.

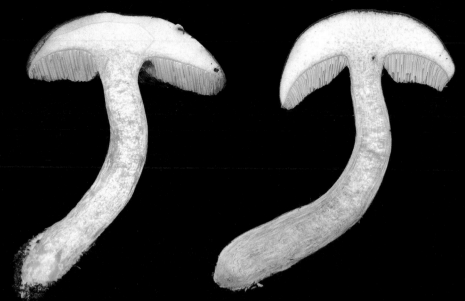

An Overlooked Edible Mushroom with a Look-alike

The Bay Bolete to the left has flesh that stains slightly blue while the Suede Bolete on the right has a more yellow flesh.

MANY MUSHROOM PICKERS pass by the Bay Bolete, not realizing that it is an excellent edible mushroom. It has a firm flesh and is reminiscent of Porcini in its taste but the species no longer belong to the genus *Boletus*; instead, it's been discovered that it belongs to the related genus *Xerocomus*.

The Bay Bolete grows in cooler coniferous forests across the Northeast and Central United States and into adjacent Canada. It's a mushroom that grows in coniferous forests but can also be found in deciduous ones. The fruiting bodies can appear as early as July and sometimes late into fall, as these mushrooms can tolerate the cold. Their quality is also better once the insect season has passed.

The classic look-alike mushroom is the Suede Bolete, which is edible but is not considered as good as the Bay Bolete.

Both species are mycorrhizal mushrooms that can also sometimes get their nourishment through breaking down dead, organic matter.

Bay Bolete *Xerocomus badius*

FRUITING BODY: Usually smaller than the ceps, which include Porcini, Summer Cep, and Pine Bolete.

CAP: 2–6 inches (5–15 cm) wide. Dark chestnut brown in color. At first ovate, then convex to flat. Starts off downy, then turns smooth and shiny and sticky when the weather is damp.

PORES: Glowing yellow on the younger specimens, then porous with hints of green. Turns blue when bruised.

STIPE: Approximately 3–5 inches (8–12 cm) tall. Brown and streaky and lighter in color than the cap. Equal thickness throughout, usually bent with a slim, pointy base.

FLESH: Pale yellow. Turns slightly blue when cut. The flesh of the stipe is surprisingly tough.

SPORE PRINT: Olive brown.

RANGE AND HABITAT: Common across North Central and Northeastern US and adjacent Canada, and as far south as North Carolina. Thrives in acidic, coniferous woods, especially with pine trees.

SMELL AND TASTE: Mild and pleasant. When cooked is reminiscent of Porcini.

PICKING AND CLEANING: In contrast to other ceps it doesn't succumb to many insect infestations, especially later in the season. Always remove the tubes.

PREPARING AND STORING: Can be enjoyed fried or stewed. When the mushroom is cooked, the blue tint in the flesh disappears and turns light yellow. It can be dried.

LOOK-ALIKE MUSHROOMS: A close relative of the Bay Bolete is the Suede Bolete, *Xerocomus subtomentosus*, which is really several species and they are all edible. Some of the distinguishing features are the velvet caps that can change from light chocolate brown to olive brown, as well as the bright yellow pores on the young specimens.

One of the species within the group *Xerocomus* (probably *Xerocomus ferrugineus*). The Bay Bolete's cap is usually darker than the Suede Bolete's.

SLIPPERY JACK

◁ Fully mature specimens are not as sticky. The caps are golden brown with violet streaks and look silky and shiny in drier weather.

◻ Young Slippery Jacks are covered by a thick, slimy coat.

THE RAYS FROM the morning sun reach into the chilly corners of the forest and wake up an insect or two. It's September and most of the plant life is winding down—but not the mushrooms. They are entering their most busy phase and are producing lots of fruiting bodies that appear above the surface of the ground.

A path tempts us further on into the forest and soon a host of streaky mushroom caps appear. It's the beautiful Slippery Jack. A delicious-tasting edible mushroom that is easy to recognize by its sticky cap, its ring, and the yellow pores.

We pick a few fruiting bodies; the caps stick to us and you can just glimpse the buttery yellow pores where the ring is coming loose at the cap margin. We remove the pore surface, which has a texture like foam.

The cap flesh is fairly firm and the flesh of the stipe is hard—just like with all ceps. The cap surface might cause stomach problems in a few people, but it's easily removed.

More goodies fill our basket and it won't be long before we get to enjoy them.

YELLOW PIECES OF Slippery Jack stand out in our pan, which is a potpourri of Porcini, Chanterelles, and Orange Birch Bolete. The butter gently simmers away, spreading a heavenly aroma all through the house.

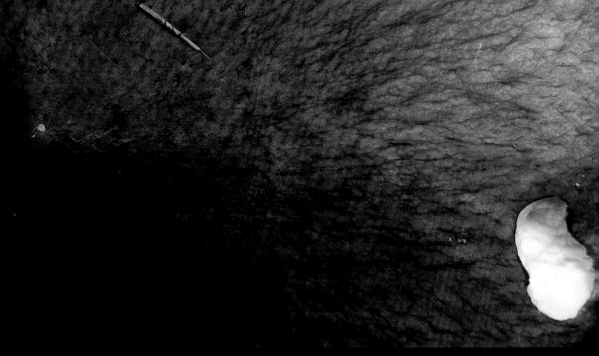

The Slippery Jack's cap often has a pointed center. The violet streaks are dried remnants of slime.

A Cep That Likes Pine Trees

THE SLIPPERY JACK belongs to the genus *Suillus* members of which often have a slimy cap. This group develops a mycorrhizal symbiosis with various coniferous trees. They are all edible, but the Slippery Jack is considered the best edible mushroom of this group.

Recent research has shown that the Slippery Jack is more closely related to the Slimy Spike Cap, *Gomphidius glutinosus* (page 193) than to ceps.

It has also been found that some species of *Suillus* can cause stomach upset in some people, so you should taste only a small portion the first time you eat one.

North America has double the number of species of *Suillus* that are found in Europe, as Europe does not have as many species of coniferous trees. In the United States, the Slippery Jack grows alone or in groups and is usually found under Scots Pine. Sometimes you find groups where the caps have grown together.

The season for Slippery Jacks tends to be late summer into the late fall and even after the frost appears.

The cap flesh is yellower than the stipe's.

Slippery Jack *Suillus luteus*

FRUITING BODY: Smaller than the Porcini. When young, the cap (pin head) and half the stipe is surrounded by a thick layer of slime.

CAP: 2–6 inches (5–15 cm) wide. Dark to golden brown in color, and young specimens are darker. When damp, it is covered by a layer of slime that becomes less sticky with age. A shiny surface when dry, with dark streaks radiating from the center. It is bumpy at first, and then spread out with a slightly pointed center.

PORES: Covered by a white veil to start with that comes loose from the margin as the cap grows. The remnants of the veil create a ring around the stipe. The color of the small, round pores is light yellow and becomes a more dirty yellow with age.

STIPE: 1–4 inches (3–10 cm) tall and yellow with dark grains above the ring. At the bottom, white to light yellow and sometimes brown with hints of purple against the thicker base.

FLESH: The light yellow flesh is firm at first and softens with age. The flesh of the stipe is lighter and firmer than the cap.

RING: White at first, then dries onto the stipe and eventually darkens from the spores that drop down from the pores. At times the ring will fall off.

SPORE PRINT: Light yellow to brown.

RANGE AND HABITAT: Grows widespread across much of North America. Thrives with pine trees, especially Scots Pine on sandy, lean soil. Grows during late summer to late fall.

SMELL AND TASTE: Pleasant and when cooked, mildly nutty.

PICKING AND CLEANING: Avoid rain. Cut the stipe just below the ring and remove the sticky remnants from the stipe. Remove the cap cover and the pore surface.

COOKING AND PREPARING: Mixes well with other mushrooms. Fry thoroughly or stew. Dry in thin slices.

LOOK-ALIKE MUSHROOMS: Weeping Bolete, *Suillus granulatus* (page 41), and the Larch Bolete, *Suillus grevillei*, are also tasty edible mushrooms. The Larch Bolete grows with larch trees. The flesh is yellow and turns brown when bruised.

The Larch Bolete's cap is lighter than the Slippery Jack's.

WEEPING BOLETE

IT'S THE END of August, and we are on a mushroom course in southern Sweden. When we're out picking mushrooms during the afternoon seminar, the summer heat makes a welcome comeback. We find ourselves in an interesting little pine forest with a view across the Skagerrak Strait, where many mushrooms are jostling for attention.

"The Weeping Bolete tastes good," one of the participants exclaims and seems more interested in the frying pan than in the microscope that awaits us back in the classroom.

Long strands of grass are stuck onto the tightly-packed, sticky mushroom caps. They really look like Slippery Jacks and we examine the underside of one of them. The stipe hasn't got a ring and we see drops hanging down just like shiny pearls under the pore surface. The area on the stipe by the cap is covered in small, light grains, which has given this mushroom its Swedish name "Grainy Bolete."

A Bolete That Has Been Wrongly Overlooked

Divided Weeping Boletes of varying ages. When mature the pores grow partially down the stipe.

THE WEEPING BOLETE, or False Slippery Jack as it has also been known, has been overshadowed by its popular look-alike the Slippery Jack. The Weeping Bolete is just as delicious, as well as having a firmer flesh.

Just like the Slippery Jack, the Weeping Bolete belongs to the genus *Suillus*, which have sticky caps, depending on the weather. It thrives with pine and prefers lime-rich soil. This species grows from late June to late fall, in pine forests across cooler regions of the United States and into Canada.

The safest way to recognize these mushrooms is by the drops that hang under the dense pore surface in damp weather, mainly on the young mushrooms. Just under the cap, on the top part of the stipe, you can find small, light grains that darken with age. The species usually grows in large groups along country lanes.

Weeping Bolete *Suillus granulatus*

FRUITING BODY: Like the Slippery Jack but lacks a ring on the stipe.

CAP: 2–4 inches (5–10 cm) wide. Starts off ovate and then wide and convex. The color varies from light to red/brown. It's nearly always lighter than the Slippery Jack and also has a layer of slime on the cap that is stickier when it is damp.

PORES: Light yellow and discharges a white liquid that hangs like drops under the dense pore openings that become larger on the older mushrooms when the liquid becomes grayer.

STIPE: 1 1/2–4 inches (4–10 cm) tall and without a ring. White to light yellow with small, white grains that are darker in the area nearest the cap. Below this, white to light yellow, sometimes brown with hints of purple against the thicker base. The flesh of the stipe is lighter and firmer than the cap. At the base of the stipe a brown patchy area often appears in the older mushrooms.

FLESH: White and yellow mix. Firmer than the Slippery Jack and gets softer with age.

SPORE PRINT: Golden brown.

RANGE AND HABITAT: Frequent in pine forests across central and Northern United States and into Canada. Thrives with pine in lime-rich soil and usually grows along country lanes. Can be found from early summer to late fall and usually appears in larger groups.

SMELL AND TASTE: Pleasant and mild.

PICKING AND CLEANING: Avoid picking these mushrooms in wet weather as the cap is very sticky. Only use young, firm mushrooms. Cut off the stipe, remove the skin from the cap and the pore surface.

PREPARING AND STORING: Same as Slippery Jack on page 37.

LOOK-ALIKE MUSHROOMS: It is common to mistake this for other safe *Suillus* such as the Slippery Jack, *Suillus luteus* (page 41), Larch Bolete, *Suillus grevillei* (page 37), and the Jersey Cow Mushroom, *Suillus bovines* (page 47).

As with all species of *Suillus*, you need to remove the skin of the cap after picking. This photo shows a cap half stripped from a Slippery Jack.

VELVET BOLETE

The Velvet Bolete thrives in the pine forest, amongst lichen, moss, and heather.

THE SEPTEMBER AIR *is all around us and an early morning rain has released the scents of the forest. A noisy raven flies high across the top of the trees and the mushroom season has reached its peak.*

We move across lean, earthy ground with stone surfaces covered by moss and lichen. The surface is teeming with Velvet Bolete, a mushroom that has climbed high on our list of favorites lately.

The sand-colored cap surface is still slightly damp but does not feel sticky. On the cap we find some small, grainy granules and tufts along the edges.

The pores have small, round openings and are cinnamon brown and softer on the larger mushrooms. The largest boletes are left as they are good for spreading the spores. It's the smaller fruiting bodies with firmer flesh that

◁ The sand-colored cap is
 covered by small grains
 and a light down.

are the best for taste and consistency. The chocolate-colored pore surface on the smallest mushrooms is usually hard enough to not have to be removed.

The stipe is smooth and firm and comes in different shades of yellow with clearer colors than the cap.

The Velvet Bolete has a special smell. Flowery and sharp, like a mixture between goat cheese and Evening Stock.

When divided the flesh is white to light yellow and slightly orange at the base of the stipe. Soon the cut area starts to shift to bluey green which is a good indication.

IT'S A GOOD HARVEST. *After finely cleaning the mushroom, it is laid out to dry in thin slices. When the parchment-like pieces are soaked, sautéed in butter, and then cooked in cream, they taste phenomenal.*

Taste-wise it is a good substitute for the Gyromitra esculenta, which is classified as one of today's most dangerous poisonous mushrooms (page 210).

◁ The flesh of the Velvet Bolete has shades of orange and goes slightly blue. Being smaller and firmer is no guarantee that they haven't been infested and a fully mature mushroom can be just as good.

▯ The Velvet Bolete tastes best dried.

A Reappraised Bolete with a Freckled Cap

THIS SPECIES GROWS in deciduous forests in the northern hemisphere except North America and Eastern Asia. In the north the Velvet Bolete grows from summer to late fall in sandy, nutrient-poor soil and it thrives with pine trees.

The Velvet Bolete belongs to the group of boletes that has a removable cap skin although it distinguishes itself from the Weeping Bolete and the Larch Bolete in that the cap is not sticky. The cap is matte and downy when dry, has small freckles and is also known as the Freckly Tube Bolete in Sweden.

The mushroom was previously considered to be only an adequate edible mushroom and it can get quite slimy when cooked from fresh.

AS DRYING HAS become a lot more popular it has been discovered that when cooked, dried Velvet Bolete has a much tastier concentrated flavor than when cooked from fresh. The consistency is also more pleasant and taste wise it can be compared to the False Morel which is now classed as a poisonous mushroom.

The Velvet Bolete can't be confused with any poisonous mushroom. The nearest false mushroom is the Jersey Cow Mushroom that is edible but not really of interest for cooking.

ONE OF SWEDEN's legends in the mushroom world is Bengt Cortin and he had a few unusual tips on how Velvet Bolete should be enjoyed as a dessert:

> If one has considered a three course meal made up of mushrooms you can get an original and great tasting dessert with a candied Velvet Bolete. The mushrooms are melted into a thick sugary syrup and are then placed out to harden.

From 'GODA MATSVAMPAR' (GOOD EDIBLE MUSHROOMS)

A perfect example of the Velvet Bolete

Velvet Bolete *Suillus variegatus*

FRUITING BODY: Just as big as the Slippery Jack and Weeping Bolete and does not have a sticky cap.

CAP: 2-6 inches (5-15 cm) wide. Sandy to rusty brown in color. At first ovate and then convex. The skin is easily removed. It has a freckly, downy surface in dry weather.

PORES: Dense. At first in shades of milk chocolate, then olive colored to cinnamon brown with small round openings that turn bluey green when pressed.

STIPE: 2-4 inches (2-3 cm) thick. Smooth and firm. Mottled yellow to a brown yellow and thicker at the base of the stipe.

FLESH: Pale yellow, often with shades of orange, especially at the base of the foot. Turns blue when cut and thumbed.

SPORE PRINT: Brown.

RANGE AND HABITAT: All over Sweden, often in vast amounts. Grows from July to October in lean pine forests with flat areas of rock and thrives with lichen, heather, and lingonberry bushes.

SMELL AND TASTE: Spicy and tart. When cooked from fresh it does not have as much flavor as when dried.

PICKING AND CLEANING: Usually necessary to remove the pores, except on the smallest mushrooms. It is rarely infested by insects or slugs.

COOKING AND STORING: The Velvet Bolete tastes great if it is cooked from dried. Soak the mushroom for at least 20 minutes. Allow the mushroom to sweat over a medium heat in a pan without any fat. When most of the moisture is gone, add a knob of butter and fry at a high heat, then lower the heat and add cream. Let it simmer for about 10 minutes and you will get a luxurious sauce resembling that of the False Morel.

LOOK-ALIKE MUSHROOMS: Weeping Bolete, *Suillus granulatus*, is a good edible mushroom (see page 39). Jersey Cow Mushroom, *Suillus bovinus*, has a sticky cap when the weather is wet and grows in the same type of forest as the Velvet Bolete. It is often found in large groups. It has yellow/pink flesh and is edible but not as tasty.

The Jersey Cow Mushroom is darker and has bigger pore openings than the Velvet Bolete; in addition the skin of the cap is smooth.

ORANGE BIRCH BOLETE

◁ A young but well developed Orange Birch Bolete where the black on the stipe has not yet developed into tufts.

The Orange Birch Bolete's flesh often gets clear blue patches at the base of the stipe.

Even the larger mushrooms can be nice and firm if they have grown fast enough.

NATURE HAS BEEN doused with a much needed shower and the glistening pearls of water are now slowly evaporating in the sun.

It's the middle of the summer and the greenery has a deep tint to it. Among the blueberry bushes appear shiny purple dots, and in about a week these berries will be ready for the picking—but today we are hunting mushrooms!

Soon an extremely orange mushroom cap appears. The stipe is covered with black tufts and it looks "unshaved," which is typical of the genus Leccinum.

The cap is still sticky from the rain but firm, and it must have grown this big in no time at all. The skin of the cap is overgrown and hangs over the cap margin like a valance.

We divide the mushroom and cut off the stipe. A clear blue patch spreads slowly where the stipe has been cut and the color toward the cap has a more purplish gray hue.

Everything fits—this is an Orange Birch Bolete.

Safe Boletes That Oxidize

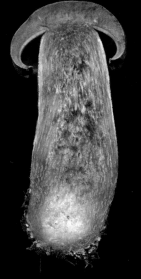

IF YOU'RE A beginner mushroom picker, the genus *Leccinum* is great to start off with. These are completely safe, easy to recognize due to the tufts on the stipe, and they are tasty edible mushrooms.

In Scandinavia, there are fourteen types of *Leccinum* and these are mycorrhizal mushrooms that are bound in symbiosis to different types of trees, for example spruce, pine, and birch trees.

When cut, the flesh reacts with oxygen and changes color. In the old days, this was thought to indicate that the mushroom was inedible; however today we know better and *Leccinum* are considered good mushrooms for eating. They do, however, need to be cooked thoroughly as they can cause nausea.

ORANGE BIRCH BOLETE grows alongside birch trees. It's the species that has the largest fruiting body of all the *Leccinum* mushrooms and is also considered the tastiest.

As with all boletes, if the weather is suitably warm and rainy the Orange Birch Bolete grows quickly. The fruiting bodies have a short life span and can quickly be attacked by insects and slugs, so it's worth checking out your mushroom picking spots regularly.

The flesh is noticeably tough and even when cooked it retains much of its elasticity and crispiness. The mushroom becomes almost black when cooked, but it tastes great.

After cutting or bruising the mushroom, the flesh reacts, oxidizes, and turns darker, from pinkish purple to blue-green and sometimes to brownish-black.

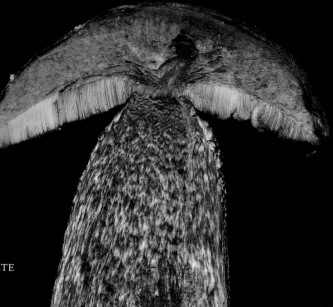

Orange Birch Bolete *Leccinum versipelle*

FRUITING BODY: Our largest species of *Leccinum*, it can weigh up to 3.3 lbs (1.5 kg).

CAP: 4–12 inches (10–30 cm) wide. Clear orange or red brown. To start with, the cap is round, then oval and finally convex. Grainy or smooth when dry, and sticky when damp. The cap skin hangs down over the margin.

PORES: First gray and hard, then lighter and spongier. Turns dark when bruised.

STIPE: 4–8 inches (10–20 cm) tall and often thicker at the base. Black tufts on a gray-white base that get sparser with age.

FLESH: White. Reacts with oxygen and darkens to a pink and blue-green shade, sometimes with a hint of clear blue at the bottom of the stipe. Gradually turns more gray-black. The cap and top part of the stipe are firm and the consistency is more fibrous towards the base of the stipe.

SPORE PRINT: Brown.

RANGE AND HABITAT: In North America, this species is often known as *Leccinum testaceoscabrum*, and please note that it has been noted to cause nausea and vomiting in some individuals. It is seen in cooler regions and is abundant in Canada and Alaska where it appears alone or in smaller groups from summer to late fall. Grows in mixed woods and fields and thrives with birch trees. Some years it can grow in vast numbers. It can also be found above the Arctic Circle.

SMELL AND TASTE: Very mild.

PICKING AND CLEANING: Roughly clean after picking. As a rule, remove the pores.

PREPARING AND STORING: Good to mix with other mushrooms. Tastes best when freshly prepared. The mushroom pieces stay firm even after cooking and turn almost completely black. It can be dried in thin slices.

LOOK-ALIKE MUSHROOMS: Red-capped Scaber Stalk, *Leccinum aurantiacum* (page 53), and Birch Bolete, *Leccinum scabrum* (page 57).

The Red-capped Scaber Stalk's cap is a clearer orange color.

Birch Bolete is smaller in size.

RED-CAPPED SCABER STALK

The flesh of the mushroom reacts with oxygen and turns a brown-purple color at first, then black.

WE CATCH A glimpse of several colorful mushroom caps and we throw the car into reverse. Down a tree-covered hill, on the edge of an open field, a whole row of beautiful mushrooms appear. With bright orange caps and surrounded by Aspen trees, it has to be a type of scaber stalk, because there are several.

We pick the youngest specimen that actually has the thickest stipe and feels slimy and slippery. The white tufts have just started to turn brown. I tap the sticky cap and it feels good and firm. The skin on the older caps are not as colorful and you can see that the flesh of the cap is not as elastic anymore. The stipes, however, have a firmer flesh and can be used if the insects haven't got there first.

Faithful to its Host Tree

THE RED-CAPPED SCABER Stalk is a type of *Leccinum* and, just like the Orange Birch Bolete, is an excellent edible mushroom. Some notable distinguishing features are the orange color on the cap and the young, white tufts on the stipe that will eventually turn orange, then brown.

This mushroom only grows with poplars, such as the Aspen. The two types of scaber stalks with the most orange caps are *Leccinum aurantiacum* and *Leccinum albostipitatum* and they are hard to distinguish between. The mushrooms of the *Leccinum* genus, typically have one or several types of trees to which they are bound. The Red-capped Scaber Stalk has several similar "tree cousins" that are all good edible mushrooms.

The white tufts on the stipe eventually turn brown as opposed to the Orange Birch Bolete which always has black tufts.

Red-capped Scaber Stalk *Leccinum aurantiacum*

FRUITING BODY: Doesn't grow as large or as thick as the Orange Birch Bolete.

CAP: 2–8 inches (5–20 cm) wide. At first round like a ball, then ovate in shape. The skin is a bright orange. It is sticky when damp and has, just like the Orange Birch Bolete, a larger skin that hangs down or is tucked under the margin of the cap.

PORES: White and turns brownish-purple when bruised.

STIPE: 4–8 inches (10–20 cm) tall. The tufts are white at first and then turn brown.

FLESH: White and firm. Reacts with oxygen and when bruised turns brownish-purple in color.

SPORE PRINT: Olive brown.

RANGE AND HABITAT: All over North America, wherever Aspen trees occur, between July and October.

SMELL AND TASTE: Mild and pleasant.

PICKING, CLEANING, PREPARING, AND STORING: See Orange Birch Bolete on page 57.

LOOK-ALIKE MUSHROOMS: Orange Birch Bolete, *Leccinum versipelle* (page 49) and Birch Bolete, *Leccinum scabrum* (page 57).

Maybe it's a Red-capped Scaber Stalk? Even experts can have difficulty distinguishing between some of the mushrooms in the *Leccinum* genus. Luckily, in North America, none are seriously toxic, although a couple may cause upset stomachs.

BIRCH BOLETE

The Birch Bolete is a common edible mushroom in cooler forests where birch trees occur in numbers. It is seen fruiting from early summer to late fall.

WE'RE NOT HAVING much luck and so far we've only found a few small Chanterelles.

But two mushrooms appear on a path in the forest, and our hopes are raised. The stipes are slim and the two small, black tufts reveal them to be Birch Boletes. The light brown caps and the light pore surfaces are also a sure sign.

The cap sways slightly on the thin stipe when we pick it. Unfortunately, the fruiting body is too mature, the flesh of the cap is spongy and would just turn to mush in the pan, and the fibrous stipe has turned woody.

Suddenly, a dull thudding vibrates the ground and ten yards away something appears from the edge of the forest—it's a wild boar piglet. It dashes across the path, followed by two sturdy sows.

So this lackluster mushroom picking trip was memorable after all!

The soft, spongy flesh of the cap is different from the flesh of the stipe, which is firmer and more fibrous.

A *Leccinum* on a Global Scale

THE BIRCH BOLETE, also known as the Scaber Stalk, lives alongside birch trees. Across North America, it is seen in mountainous regions in the south and among birch stands across the cooler boreal forests of the northern United States and Canada.

Boletes can be found all over the northern hemisphere in temperate zones and also above the Arctic circle in Greenland and Alaska. The mushrooms can even be found in South America, as they have been imported with birch trees.

Most of the *Leccinums* have flesh that oxidizes and changes color when cut; however, the Birch Bolete does not react in this way.

There are several species that are closely related to the Birch Bolete that have not been researched as of yet, but they are all equally tasty edible mushrooms.

Birch Bolete *Leccinum scabrum*

FRUITING BODY: Not as thick as the Orange Birch Bolete, more like the Red-capped Scaber Stalk. However, it does not change color when bruised or cut.

CAP: 2–6 inches (5–15 cm) wide and convex. Turns from beige, to light brown, to gray-brown in color. Is smooth and somewhat sticky when damp, and matte and downy when dry.

PORES: Puffy. White at first but then turning gray. Often have brown patches and the pores do not reach the stipe.

STIPE: 3–8 inches (7–20 cm) tall. It has small gray, black, or brown tufts that create

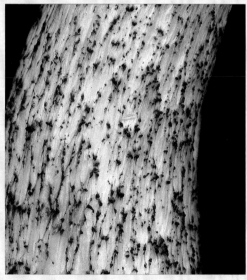

A close up of the Birch Bolete stipe, showing the tufts that create a pattern of vertical lines.

vertical lines along the whole stipe. It is thinner towards the base of the cap.

FLESH: The flesh of the cap is at first firm but quickly turns soft and watery. The flesh of the stipe is fibrous and with age turns woody and tough.

SPORE PRINT: Light brown.

RANGE AND HABITAT: All over cooler regions of North America together with the birch tree. Often grows in forests with a mixture of trees and even along the edge of fields. Grows alone or in groups, from summer to late fall.

SMELL AND TASTE: Mild and pleasant.

PICKING AND CLEANING: Only pick young, firm mushrooms. Generally it's a good idea to remove the pores.

PREPARING AND STORING: Pairs well with other mushrooms, as it doesn't have a great deal of taste. The flesh darkens when cooked.

LOOK-ALIKE MUSHROOMS: Red-capped Scaber Stalk, *Leccinum aurantiacum* (page 53). *Leccinum pseudoscabrum* grows with hazel and beech. The cap is dry and often cracked. At first light beige, then darker and finally black. All these mushrooms are edible.

CHANTERELLE

◁ The fruiting body
of a Chanterelle
lasts about two
months. A Porcini,
by contrast, only
lasts a few days.

▯ Early Chanterelles
pop up along the
side of the road and
animal trails, as
well as in fields and
forests.

WE'RE CHECKING ON *our mushroom patches and wondering if the
Chanterelles that were only tiny buttons a few weeks ago are still growing
in the field.*

Oh yes, no one else has got in before us!

*It's an old field where scythes and animals have shaped the vegetation
that grows under birch, hazel, and ash. A month ago, sweet wild strawberries were growing here, but now there are new delicacies on the menu.*

IN JUST ABOUT *three weeks "our" mushrooms have become ready to pick.
Chanterelles grow slowly and need lots of rain; if it's too dry they will stop
growing.*

*The caps create a beautiful arch down in the valley and they are per-
fectly camouflaged among a few yellow leaves. We pick them, trying to avoid
damaging the smallest mushrooms, which we leave to grow a bit bigger.*

*A car approaches and we start walking toward a group of flowers to
avoid revealing our secret mushroom spot until the intruder has passed
us by. Before these delicacies land in our basket, I take a deep breath and*

CHANTERELLE 63

inhale the smell of the Chanterelles. They smell fantastic, like a full-bodied aftershave. Discreetly flowery but also tough and peppery; it evokes an early childhood memory.

Barely at eye level with the iron stovetop, the golden mountain of Chanterelles towers up out of the pan. Butter has been liberally added and the pan spits and sputters, releasing steam and the most amazing smell of nutty toffee around the country cottage kitchen. This was where my lifelong affair with Chanterelles began!

A Popular Edible Mushroom

THIS IS OUR number one favorite mushroom. Returning home with a basket full of Golden Chanterelles that smell heavenly is every mushroom picker's dream, and one that does often become reality. The Golden Chanterelle ticks every box for mushroom pickers and is in our kitchen to stay. It's easy to recognize, has no dangerous look-alikes linked to it, and is usually free from insect infestation and, of course, it tastes fantastic!

It grows in both coniferous and deciduous forests and is common all over cool forests in North America. The species thrives in a variety of terrains and has a long growth period, usually from the end of June to late fall. The first Chanterelles of the season appear along forest paths, road sides, animal trails, and fields. Towards the end of July they start to grow in the forest, which should be sparse, older, and without too much nitrogen in the soil. Chanterelles grow slowly and require a lot of rain. At the end of June you can usually see clusters of Chanterelle pinheads; unfortunately, they

The ridges on a Chanterelle reach from the stipe up to the edge of the cap and are often connected to lower, horizontal ridges.

Cantharellus amethysteus has a scaly cap.

Cantharellus melanoxeros darkens when dry.

It's fairly easy to train a dog to find chanterelles, because chanterelles have a strong aroma.

often don't grow any bigger when the warmer and drier weather appears.

THERE ARE APPROXIMATELY seventy known varieties of chanterelle in the world. The most common and well known across North America, the Golden Chanterelle, grows on four continents, mainly in the northern hemisphere. *Cantharellus pallens* is fairly common in some hardwood forests and just as good as the usual chanterelles. It usually appears in early summer and grows specifically in leafy woods in the northern United States and Canada (page 69).

Cantharellus amethysteus is common in central Europe but does not occur in North America. It looks like a regular chanterelle but the mature mushrooms have small scales in concentric circles that are pale violet to brown.

Chanterelle *Cantharellus cibarius*

FRUITING BODY: Cone shaped and not as thick as *Cantharellus pallens*.

CAP: 1–5 inches (3–12 cm) wide. Light yellow, bright yellow to orange in color. When young looks like a button and then turns more cone-like or vase-shaped. The margin stays folded in for a long time and often becomes wavy and uneven at maturity.

RIDGES: Fork-shaped toward the cap and reach far down the stipe. They are irregular and wrinkly.

STIPE: 1–6 inches (3–15 cm) tall. The color is similar to the cap or a lighter yellow to almost white.

FLESH: Pale yellow and firm.

SPORE PRINT: Pale yellow.

RANGE AND HABITAT: Found widespread across North America and the UK. It thrives in both coniferous and deciduous forests. Usually grows in scattered groups or arches from late June to late fall when the weather is mild.

SMELL AND TASTE: Pleasant and reminiscent of dried apricots.

PICKING AND CLEANING: The Chanterelle is rarely insect infested. Cut off the earthy stipe and brush off any dirt. Sometimes the fruiting bodies are extra dirty and need to be rinsed under running water. Divide the mushroom.

PREPARING AND STORING: Has a spicy taste and works well in stews or fried with butter. Mix with other mild-tasting mushrooms. It does not lend itself to drying, as it will turn tough and bitter. The best thing to do is to soak and then freeze the Chanterelles if you plan to store them (page 4).

LOOK-ALIKE MUSHROOMS: *Cantharellus pallens*, tastes like a regular chanterelle (page 69). *Hygrophoropsis aurantiaca*, the False Chanterelle has a yellow orange cap, darker stipe, brilliant orange gils, and is smaller when mature. It is not dangerous but is of no interest as an edible mushroom.

The Chanterelle is yellow.

The False Chanterelle is smaller and more orange.

CANTHARELLUS PALLENS

The underside is often a beautiful lemon or yolk color.

LAST SEASON'S BEECHNUT *husks crunch under our feet as we walk. It's Midsummer Night's Eve and no, we are not collecting seven different types of flowers to lay under our pillow—even if Swedish tradition calls for it. We're in a beech forest with several well-tramped paths where deer have walked. There aren't many herbs under the thick veil of leaves, but our chanterelle spot at the edge of the forest often pays out at this time of year.*

It took us a few years to realize that it wasn't the regular Golden Chanterelle but a different species altogether, the cantharellus pallens.

We can see them from miles away, chubby with creamy yellow caps; the salmon supper we have planned for later will be a great accompaniment!

A Mushroom That Demands Nutrients and Leafy Trees

Three young *Cantharellus pallens* with thick stipes. The caps can be creamy white, pale yellow, or, as in this case, a more pink or apricot color. The stipe and ridges have flecks of lemon yellow and brown patches will appear when bruised.

IT WASN'T UNTIL 1959 that *Cantherellus pallens* was classed as its own species. Previously, it was known as the "summer chante-relle," as it appeared early in the season—usually around June if the early summer rain was heavy enough. It was also thought that the Chanterelle grew differently in deciduous forests.

The mycologist Waldemar Bülow describes their appearance in his 1916 book *Mushrooms for the Home and School (Svampar för hem och skola)* "It has a pale yellow shape with a thick cap and a stipe that is about 1 inch (2 cm) thick."

Cantharellus pallens, or *C. cibarius* var. *pallidifolius*, as it is known in North America, can be found growing with oak, beech, and other hardwood trees across the cooler regions of North America. It pre-fers nutrient-rich soil but also soil with lime and birch. It used to be easy to find in pastures rife with leafy trees; however, these days it is much harder because of modern farming techniques.

Those in the know are of the opinion that the *Cantherellus pallens* is better than the normal Golden Chanterelle.

Cantharellus pallens (aka. C. cibarius var. pallidifolius)

FRUITING BODY: Short and chubby and turns a patchy brown when bruised.

CAP: 2–5 inches (5–12 cm) wide. At first it has irregular humps with a rolled-in margin and then turns flatter, sometimes with a sunken middle. It shifts in color from creamy white to light yellow and even shades of pink and apricot. It is often lighter at the margin.

RIDGES: Fork-shaped toward the cap and decurrent. They are irregular and wrinkled and sometimes crimped in the middle. Pale to butter- or lemon-yellow in color.

STIPE: 1–3 inches (3–8 cm) tall and lighter than the ridges.

FLESH: Yellow white and compact.

SPORE PRINT: Pale yellow.

RANGE AND HABITAT: Appears in leafy forests or fields rich in lime in northern areas with beech, oak, and birch, but even with lime and birch. Grows in groups or bouquets.

SMELL AND TASTE: Pleasant, like the Chanterelle.

PICKING, CLEANING, PREPARING, AND STORING: See Chanterelle, page 61.

LOOK-ALIKE MUSHROOMS: Chanterelle, *Cantharellus cibarius* (page 61), and False Chanterelle, *Hygrophoropsis aurantiaca* (page 67).

Divided fruiting bodies. The *Cantharellus pallens* to the left is thicker and fleshier than the Chanterelle to the right.

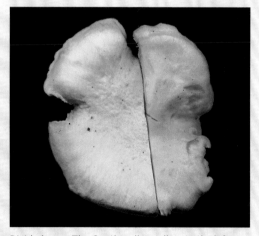

Divided caps. The *Cantharellus pallens* to the left has a lighter cap that shifts in different shades. The Chanterelle to the right is a darker yellow.

TRUMPET CHANTERELLE

LONG, GRAY CLOUDS cover the sky. It's October and yellow birch and aspen leaves slowly sail from the trees onto the ground. Dark pine trees silently watch over the forest and there is a chill in the air that nips at our faces.

Overgrown tire tracks lead us into the old, moss-covered forest. There is no wind, nor any mosquitoes, just a solitary bird that tweets and then flies off. The trees prevent the daylight from shining through, but after a short time our eyes grow accustomed to the gloom. We are on the hunt for autumn's easiest mushroom target—the Trumpet Chanterelle—and we need to activate our "trumpet eyes" as the small, brown caps have a habit of blending into the ground.

At the foot of a tree stump we find our first group of mushrooms. The smallest ones are light brown with caps the size of shirt buttons. On the larger ones, the hollow on the cap is seen more clearly and the margins are still somewhat folded in.

The larger the mushrooms, the more cone-shaped the cap, with irregular shapes and clear ridges on the underside.

The golden brown stipes are slightly buckled and go deep into the moss. We grab them in bunches and check them carefully before placing them in the basket. A few rogue species can sometimes hide among the delicacies.

WE LEAVE ALONE the tiniest mushrooms, as well as the bigger mature ones, which can taste bitter. So too with the ones that feel soft and saturated with liquid, because these have been attacked by the frost one too many times.

Despite many chilly evenings, most of the mushrooms have pulled through. The caps are the first to take the hit from the cold and they get dark and loose or black and shriveled.

We scan the terrain and everywhere we look we see hosts of mushrooms.

The difficulty with Trumpet Chanterelles is when to stop picking them! You can easily get greedy and return home with more mushrooms than you know what to do with. Even if cleaning them is easy, every mushroom needs to be divided and examined for slugs and pine needles that may hide inside the cap. It's time-consuming and fiddly and even chanterelle lovers can find this task a bit tedious!

The Trumpet Chanterelle's ridges turn gray and become more visible with age.

Withstands the Cold and Grows Plentifully

You can even pick Trumpet Chanterelles in the snow, and they make an interesting sound when they rattle in the basket. As they are frozen, they need to be cooked and eaten straight away.

IT HAS TAKEN a while for the Trumpet Chanterelle to appear on our tables. It was in the middle of the twentieth century that this mushroom was deemed worthy to be placed higher on the edible mushroom list. Its popularity in the United States is steadily growing as people discover its high quality.

These days, the Winter Mushroom, as it is also known, is one of Sweden's most popular edible mushrooms, and if you can't make it into the forest to pick it yourself, it's readily available in the grocery store during the Trumpet Chanterelle season.

TRUMPET CHANTERELLES GROW from August to November, or even later if the snow is delayed, as they can withstand frosty nights and even a bit of snow. They prefer moss and pine and hemlock forests and grow over the Central and Eastern United States and adjacent Canada. The fruiting bodies usually appear in large quantities and are at first hard to see, but once you get accustomed to what to look for, you can usually find as many as you want.

The species can even be found in large parts of Europe in both pine and beech forests.

The caps of Trumpet Chanterelles in various colors and shapes (see the following page as well).
Sometimes they can be found grown together. The stipes are not always hollow all the way through
and have a porous flesh in the middle.

Trumpet Chanterelles come in a variety of guises, both in color and shape. The photo on the bottom left shows solid yellow Trumpet Chanterelles that lack black pigment. On the bottom right, you can see how densely the mushrooms can grow. The white patches on the caps are not a sign that the mushrooms are turning bad, they are spores that have fallen down from the mushrooms above.

Trumpet Chanterelle *Cantharellus tubaeformis*

FRUITING BODIES: Smaller than the Chanterelle, with thin flesh.

CAP: 1–2½ inches (2–6 cm) wide. Funnel-shaped with a hollow. Can be scaly. Yellow-brown to gray-brown in color and lighter when dry. The flesh is light, thin, and soft.

RIDGES: Shallow and fork-shaped. Turn from yellow to gray and get more visible with age. The margins are often irregular and wavy.

STIPE: 1½–3 inches (4–8 cm) tall. Partially or completely hollow. Irregular and buckled and often wider at the bottom. The color is lighter than the cap, yellow to brown in color. The flesh is tough and sometimes has a porous "marrow" in the middle.

SPORE PRINT: Whitish to pale yellow.

RANGE AND HABITAT: Found growing over Central and Eastern US and adjacent Canada. Thrives in pine and hemlock forests or mixed forests. Appears in groups and usually in vast quantities from August until the winter season is established.

SMELL AND TASTE: A pleasant and spicy odor. Tastes good and well-rounded when cooked.

PICKING AND CLEANING: Don't pick them in bunches without checking the mushrooms first. Other more dangerous mushrooms can grow amongst them. Clean them in the forest, as this will prevent you from picking too many.

PREPARING AND STORING: Works well in soups, sauces, and stews among other things. Excellent when dried in two or four pieces.

LOOK-ALIKE MUSHROOMS: Yellow Foot, *Cantharellus lutescens* (page 81), is an equally tasty edible mushroom. Jelly Baby, *Leotia lubrica*, and *Chrysomphalina chrysophylla*, are both safe but of no interest as edible mushrooms. There have been incidents in Europe when the poisonous Deadly Webcap, *Cortinarius rubellus*, has accidently been picked when the proper care has not been taken (page 210).

The Trumpet Chanterelle can sometimes be confused with Jelly Baby, which has a jelly-like, fleshy cap.

Chrysomphalina chrysophylla has real gills as opposed to the Trumpet Chanterelle's ridges.

Be careful when picking. Here a Webcap is growing among the Trumpet Chanterelles.

YELLOW FOOT

THIS IS JUST *the way we like it: an abundance of mushrooms.*

Along a babbling forest brook we see an army of Yellow Foot. The area is backlit to reveal these delicacies, which are otherwise hard to spot. A host of bright yellow stipes look almost luminescent against the dull colors of the forest.

The funnel-shaped caps have a more discreet color of several shades of brown. They look confusingly similar to decaying leaves and are almost impossible to spot from above. The cap margins are beautifully crinkled in irregular, wavy shapes. Close up, small scales or bumps in concentric circle patterns can be seen around the depressed center that goes into the hollow stipe.

The Yellow Foot's colorful stipe goes deep into the moss. Up by the underside of the cap, the strong orange color turns a light apricot and the ridges are hardly noticeable. The smell is strong but pleasant, flowery with a sharp fruitiness.

THE FRUITING BODIES *grow in large groups and it's quick work if you pick them in bunches. We leave the smaller ones to grow, and in a few weeks we can go on another mushroom hunt—unless the greedy moose or deer have gotten there first.*

◁ Yellow Foot grow near moving water.

◻ The Trumpet Chanterelle is the classic look-alike to the Yellow Foot, and they are equally tasty edible mushrooms.

Further up the path we see a group of Trumpet Chanterelles and this is the main look-alike mushroom to the Yellow Foot. Side by side, the difference is clear, but even if you do make a mistake, they're both good edible mushrooms.

During the mushroom season, many stores offer Trumpet Chanterelles for about $5 a pound but occasionally the basket is filled with, you guessed it, Yellow Foot.

A Bouquet of Flowers

YELLOW FOOT THRIVES in limey soil and mossy spruce forests as well as in damp environments. On the island of Gotland, off the east coast of Sweden, as well as in the Stockholm archipelago which has a similar limey soil, you can find the mushroom among lichen in pine woods.

Occasionally some Yellow Foot specimens lack one or more color pigments and these are yellow, gray, or white.

The growth period starts in August and stretches into October or until the frost starts. It's often found in large quantities in a single spot that can yield several harvests in the same season. If you want to find a new place to pick Yellow Foot, look on a map for marshland, as this mushroom thrives on the edges of this type of terrain.

YELLOW FOOT HAS a very special smell that is absolutely fascinating. Bengt Cortin and Nils Suber, authors of several books on mushrooms, liken it to wild strawberries and ripe pears. Some people find that there is a flowery smell in the room when drying these mushrooms.

The stipes are hard, hollow, irregular, and flat.

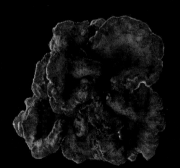
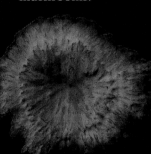
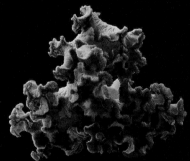

The caps can look very different to each other and grow in the strangest crinkles and wrinkles.

Yellow Foot *Cantharellus lutescens*

FRUITING BODY: Very similar to the Trumpet Chanterelle.

CAP: 1–3 inches (3–8 cm) wide. Funnel shaped and slightly bumpy, sometimes downy and mostly with a groove. The colors vary within shades of brown, from light to dark depending on the age of the mushroom and the weather. The cap margins are irregular and crinkled. The flesh is thin and soft.

RIDGES: Shallow, forked, and sometimes invisible.

STIPE: 1–3 inches (3–7 cm) tall and yellow to strong orange in color. Toward the cap a paler yellow/pink or almost white color. The flesh of the stipe is tougher than the cap and is usually hollow.

SPORE PRINT: Whitish to pale yellow.

RANGE AND HABITAT: Mainly in the cooler, more northern regions of the United States and into Canada. Usually in vast quantities near water.

SMELL AND TASTE: Strong with a mild, pleasant taste, and deep character.

PICKING AND CLEANING: If there are a lot of mushrooms, you can use scissors when picking them. The mushroom is easy to clean; you just need to remove the earthy stipe and divide the fruiting body, as there is often rubbish in the stipe.

PREPARING AND STORING: As Yellow Foot have a thin flesh, they work really well when dried. When cooked, Yellow Foot can be used alone or with other mushrooms as an accompaniment to fish and meat, or to add to sauces.

LOOK-ALIKE MUSHROOMS: Trumpet Chanterelle, *Cantharellus tubaeformis*, is an equally tasty edible mushroom and the only doppelganger (page 73). The other Yellow Foot, *Craterellus ignicolor*, is found in cool mixed forests with beech and hemlock. It has a yellow-brown cap and bright yellow stipe and is equally edible. Occasionally one might pick a Jelly Baby, *Leotia lubrica*, which is a safe but inedible mushroom. It is fleshy and jelly-like (page 79).

The Yellow Foot to the left has an orange stipe and a light yellow color underneath the cap. The Trumpet Chanterelle to the right is darker both under the cap and on the stipe.

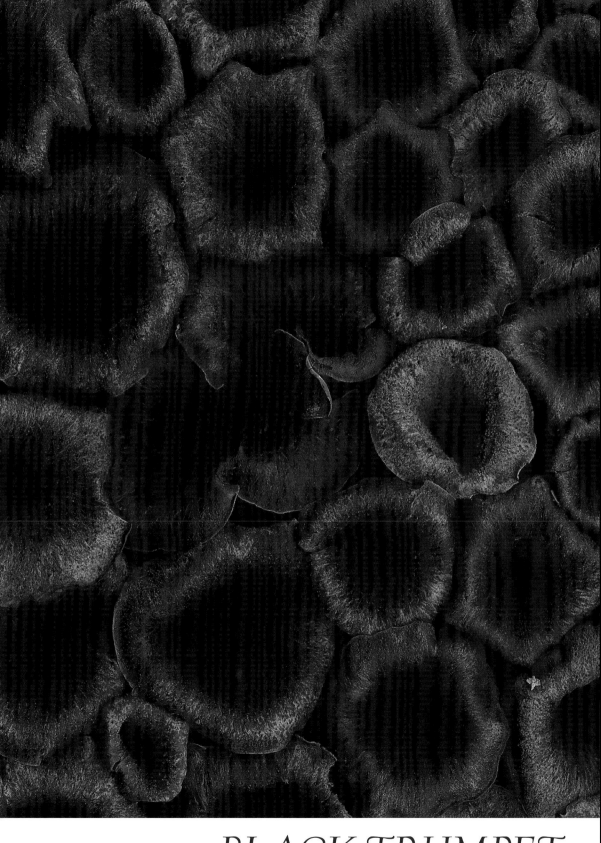

BLACK TRUMPET

The Black Trumpet is like no other mushroom, either in appearance or in taste. It's usually harvested in the same location, several times during a season.

POP! THE GIANT *Black Trumpet reluctantly lets go of its mossy hold.*

It feels like a big tulip, stiff and slightly cool and slippery. On the outside, a matte blue-gray shade shimmers and faint veins are visible. The margin of the cap is pretty and crinkled. And around the cap surface on the inside, small bumpy scales can be seen on the otherwise brownish-black flesh of the cap. The smell is unique: mild, sweetly spicy, and with a hint of perfume. The Black Trumpet is hard to beat as a delicacy and is a favorite among gourmet chefs.

I'm in my element as I survey the vast number of Black Trumpets on display.

WARM WINDS HAVE *blown in from southern Europe and it's over 68°F (20°C), despite it almost being October. Thankfully the mosquito season is a mere memory.*

This is the second time this fall that we've been to "our" Black Trumpet forest. The baby trumpets that were left to grow have gotten a lot bigger and many more have popped up, so the small basket is quickly filled. According to the mushroom guides, Black Trumpets can grow up to 6 inches (15 cm) in height and the cap can grow to 3 inches (8 cm) in width. These giants are definitely bigger. The conditions must have been perfect these past few weeks.

I CUT OFF the bottom part of the hollow stipe, which is black, and divide the mushroom, removing a few pine needles that have found their way inside.

The fruiting bodies feel lithe and have a fresh smell to them. If the flesh is soggy and smells like a rank dish cloth, they are too old.

We pick them with the understanding that there are probably many more patches of Black Trumpets in the vicinity, so we go nuts and pick everything in sight!

◁ The Black Trumpet changes color depending on the weather. The drier it is, the lighter in color. In damp weather they are black or blue-black and will stain other mushrooms in the basket.

▯ There wasn't enough space in the basket. 16 pounds (7½ kg) of Black Trumpets had to be carried home in an IKEA bag!

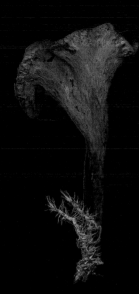

A Hard-to-Find Delicacy with a Taste of Cocoa

THE BLACK TRUMPET mushroom is one of our hottest targets in the forest. Taste-wise it tops most people's lists of favorite edible mushrooms and its rich aroma is considered to elevate a dish to heavenly heights.

If you're lucky enough to spot these "invisible" delicacies, they're easy to recognize. The Black Trumpet is unique in its appearance and has no dangerous look-alikes. It usually appears in groups and some years in vast quantities.

The fruiting bodies keep fresh for a long time in their growing spot, as opposed to boletes, which only last a few days.

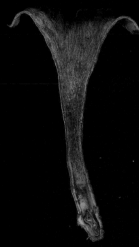

THE VARIETY IS spread across the northern hemisphere's temperate zones. It grows over much of North America in cooler forested regions.

It wasn't until the 1930s that the Black Trumpet was appreciated as an edible mushroom in Sweden, but it has been highly regarded by mushroomers in North America for well over a hundred years.

Fresh, fried Black Trumpet has a hint of fruitiness and a vaguely burnt character with a subtle aftertaste of bitter chocolate or cocoa, as well as an earthy taste.

DIETER ENDOM, CHEF AND MUSHROOM EXPERT

The outside and the inside are different in color and structure.

A dry summer is the reason why the Black Trumpet to the left is so light. The right one has had more normal weather conditions.

Black Trumpet *Craterellus cornucopioides*

FRUITING BODY: Straight and cone shaped and hollow like a trumpet. Can grow up to 6 inches (15 cm) tall and 3 inches (8 cm) wide. The colors run when wet.

OUTSIDE: This is where the spore-producing hymenium is found, which is smooth with faint veins. It is lighter than the inside, a beige gray color when dry and a glittering blue-gray when the ground is damp. At the bottom of the stipe, it is darker and brown to black in color.

INSIDE: From a warm brown to a blue-black. The cap is covered in small scales or bumps and the surface is matte and downier further down the mushroom where insects and debris often hide.

FLESH: Thin, lengthwise fibers, chewy and leathery.

SPORE PRINT: White.

RANGE AND HABITAT: Widespread across much of North America in deciduous or mixed woods with rich soil (usually beech, oak, and hemlock). It also thrives along mountainous regions with older spruce forests. Often appears in large groups and grows in clusters. The Black Trumpet has a long season and can be harvested several times during the season.

SMELL AND TASTE: A pleasant, unique, and spicy smell. When prepared it tastes delicious and is reminiscent of dark chocolate and Morels.

PICKING AND CLEANING: It is rarely infested. If the caps' margins are black and dry they can be removed. If the mushroom is partially or totally black, sloppy, and smells bad, it is too old. The darker part at the bottom of the stipe should always be removed, as it tastes bitter.

PREPARING AND STORING: It is good when fried with butter or cooked in cream for 20–30 minutes. It also makes a great accompaniment to both meat and fish and is very well suited to drying; it should be soaked for at least 20 minutes before cooking. In our experience, you shouldn't use the remaining liquid, as it tastes bitter, but this is not everyone's experience.

The dark patch on the top right smells bad and this mushroom has started to decompose.

LOOK-ALIKE MUSHROOMS: You might possibly confuse it with *Craterellus cinereus*, but this one has obvious ridges under the cap. It's edible but very rare.

WOOD HEDGEHOG

The teeth hang down like a sparsely woven mat underneath the cap.

THE "LOST WOODS" . . . this is what we call it, since we got lost here one time after the lure of mushrooms tempted us further and further away from the known terrain. We used old scouting tricks, like ant hills to the south and mossy tree trunks to the north and finally got back on the right track.

Some years the ground in the "lost woods" is teeming with Wood Hedgehog late into the fall, just like now, and an arch of creamy colored caps are visible from afar. We pass mushrooms that weren't able to cope with the frosty nights and have turned black and shriveled, their identities lost. But the Wood Hedgehogs are hardy and rarely affected.

We remove the rubbish and earth from the cold stipes and the flesh crumbles—a bit like a hard lump of fresh yeast. We scrape away the teeth to avoid the "coconut" effect in the frying pan.

The sun is setting behind the treetops and it's starting to get dark . . . and where is that path again?

Hard to Distinguish

The teeth are sharp and travel down the stipe in a soft arch. The flesh has patches that are carrot yellow.

MUSHROOM GUIDES RATE the Wood Hedgehog higher than its twin the Terracotta Hedgehog but many think they are equal in taste. Both species share so many distinguishing features that it can be hard to see which of these tooth fungi it is.

The Wood Hedgehog can almost turn red/yellow in color and the Terracotta Hedgehog can look pale. The Wood Hedgehog often grows in large groups with caps that have grown together, and will have a leaning stipe. This can also apply to the Terracotta Hedgehog, but very rarely.

You should avoid picking the larger ones as they are bitter with a soapy taste.

IN SWEDEN, WOOD Hedgehog appears all over the country in deciduous, mixed, and coniferous woods from late summer all the way through to November, if winter doesn't come too early.

Wood Hedgehog grows in Europe, North and Central America, Australia, and some parts of Asia. It's a popular edible mushroom and is sold in many parts of the world.

A pale form of the tooth fungi is common in some North American forests, *Hydnum repandum* var. *album*. It is equally edible and choice.

Wood Hedgehog *Hydnum repandum*

FRUITING BODY: Is somewhat reminiscent of the Chanterelle but with teeth instead of ridges.

CAP: 1–6 inches (3–15 cm) wide, matte and often irregularly chubby and fleshy. Convex and sunken in the middle with a wavy margin. The color varies from white to creamy yellow and chamois-colored, and sometimes even salmon pink.

STIPE: 1–3 inches (3–8 cm) tall and fleshy, sometimes at an angle and thicker at the base. White to yellow-white and often carrot yellow patches at the bottom of the stipe.

TEETH: Dense and start off white, then turn the same color as the cap. Decurrent and detach easily on the more mature mushrooms.

FLESH: White to pale yellow and dense. Breaks down easily and has rusty brown patches when handled.

SPORE PRINT: White.

RANGE AND HABITAT: Widespread across cooler regions of North America from August through November. In coniferous, mixed, or deciduous woods and often in dense groups, rings, or arches.

SMELL AND TASTE: Mild and pleasant.

PICKING AND CLEANING: Tooth fungi are often free from infestation and the flesh is quite tender, so be careful when you pick it. Young to half-mature mushrooms are the best. The big ones can be bitter and taste soapy. You can eat the teeth but they might detach in the pan and will sprinkle your meal.

PREPARING AND STORING: Works well mixed with other types of mushrooms, such as chanterelles. Not suitable for drying as the bitter taste gets concentrated when dried, but this disappears when cooked from fresh, frozen, or conserved.

FALSE MUSHROOMS: Terracotta Hedgehog, *Hydnum rufescens*, is an equally tasty edible mushroom (page 99). You might possibly mistake the Sheep Polypore, *Albatrellus ovinus,* for a Wood Hedgehog, but this has a thin pore surface instead of teeth (page 103).

Albatrellus ovinus, the Sheep Polypore, and *Albatrellus confluens* are also good edible mushrooms and can sometimes be confused with the Wood Hedgehog.

TERRACOTTA HEDGEHOG

The transition between cap and stipe is very clear.

AN ORANGE FLASH of mushroom caps can be seen under the bushy spruce. Could it be Chanterelles? A closer look shows the color to be more apricot and under the cap we see densely packed teeth; it's the Terracotta Hedgehog.

The color contrasts beautifully against the fluffy green moss. Sometimes it's hard to tell the difference between the Terracotta and Wood Hedgehogs, because they thrive in the same type of terrain and can be very similar both in color and shape.

We pick one to examine. The teeth finish sharply against the stipe, which is typical of the Terracotta Hedgehog.

Whatever the mushroom season, Terracotta and Wood Hedgehogs are faithful friends that can almost always be found. They don't mind cold, damp, or dry weather and have a long season. If you should confuse one with the other, you'll still have a great tasting edible mushroom.

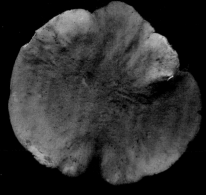
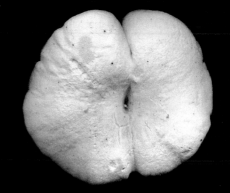

The Terracotta Hedgehog is often smaller in size with a thinner stipe. The orange cap is thinner and the teeth finish sharply at the stipe.

The Wood Hedgehog is usually bigger, thicker, and with a fleshy, white-yellow cap. The teeth go down the stipe with a soft transition.

At Least Three Varieties

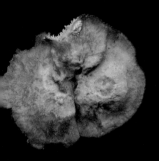

Possibly a different variety of the Terracotta Hedgehog. The color is stronger and the size smaller.

TERRACOTTA HEDGEHOG GROWS all over Sweden, so we have no trouble finding it. It likes the same type of terrain as the Wood Hedgehog but prefers slightly damper ground and grows more solitarily. It has not gained its deserved ranking among the edible mushrooms and is usually only mentioned as a look-alike mushroom to its more popular cousin the Wood Hedgehog.

In recent years, it has been found that the Terracotta Hedgehog consists of at least three different varieties. This would explain the different appearances that can be found, from the thick, densely growing fruiting bodies to the smaller, more solitary examples.

Terracotta Hedgehog *Hydnum rufescens*

FRUITING BODY: Very similar to the Wood Hedgehog.

CAP: 1–3 inches (3–7 cm) wide. Downy, thinner, and more regular than the Wood Hedgehog. Red/yellow to orange brown in color.

STIPE: 1–3 inches (2–7 cm) tall, often centered and thicker at the bottom. White to yellow-white in color, usually with flecks of orange at the bottom of the stipe.

TEETH: Dense, and they finish at the stipe. White to red/yellow and detach in the more mature mushrooms.

FLESH: White to pale yellow, sometimes carrot colored. Often delicate and more fragile than the Wood Hedgehog. Discolors when handled. The flesh of the cap and stipe are equally good.

SPORE PRINT: White.

RANGE AND HABITAT: This mushroom is uncommon, but it has been reported from the Central Appalachian Mountains in coniferous, mixed, and deciduous woods. Often grows alone or in groups but even in rings or arches.

SMELL AND TASTE: Mild and pleasant, similar to the Yellow Chanterelle when cooked.

PICKING, CLEANING, PREPARING, AND STORING: See Wood Hedgehog page 103.

LOOK-ALIKE MUSHROOMS: Wood Hedgehog, *Hydnum repandum*, is an equally tasty edible mushroom. *Hydnum umbilicatum* is a small hedgehog with a depressed cap center and is common in moist conifer woods, especially with hemlock across cooler regions of North America. *Albatrellus confluens* has a thin pore surface instead of teeth (page 105).

Avoid picking larger examples of the Terracotta Hedgehog (left). This goes for the Wood Hedgehog as well (right). They taste bitter and are more suited to be spore spreaders.

SHEEP POLYPORE

◁ An older spruce forest in Hälsingland, Sweden provides an impressive fairy ring of Sheep Polypore.

⬚ These are a good-sized mushroom to pick. The Sheep Polypore's shallow pore surface under the cap is white but it turns yellow with age.

IT'S A RARE experience to enter a forest where you haven't stepped for the last thirty years. The land still feels familiar; it's just the trees that have grown taller.

Our memories of this forest keep popping up; isn't there a place up the hill where we used to pick Sheep Polypore? We would cook the Sheep Polypore together with soy in an iron pan to make a false truffle.

Sometimes we talk about a mushroom as an individual, and in this case, the personification is justified. When you see several fruiting bodies in a limited area, most times you're actually looking at just one mushroom.

The Sheep Polypores in our old picking spot have had a long time to grow and we're not disappointed to see that they have created a magically beautiful fairy ring.

This is a happy reunion.

A Mushroom in the Food Industry

THE SHEEP POLYPORE is a great edible mushroom. It grows in large amounts and has no dangerous look-alike mushrooms. It's common in conifer forests across the northern regions of North America. The Sheep Polypore is a mycorrhizal mushroom that grows with spruce and prefers mossy, stone-filled woods that are slightly elevated, preferably in a glade.

Previously, it was common for the Sheep Polypore to be sold in stores and today it's used in the food industry as a substitute for truffles; for example, the black bits that can be found in liver pate are actually Sheep Polypore. Its firm consistency and mild taste increases the areas of use. The cap, for example, can be breaded and fried whole. The mushroom can also be ground to a mince.

Opinions on the Sheep Polypore vary as the taste is very unique (even if it is mild), and some liken it to almonds or pork.

The thin pore surface cannot be removed as is possible with boletes.

Insect larvae can be quick to attack Sheep Polypore.

The Sheep Polypore has always been important in trade. This picture, taken at the end of the 1940s, shows a family in Hälsingland, Sweden with a large harvest of Sheep Polypore. In this case the mushrooms are being shipped to England.   From *TASTY EDIBLE MUSHROOMS* (Goda Matsvampar)

Sheep Polypore *Albatrellus ovinus*

CAP: 3–8 inches (7–20 cm) wide. When young it is at first matte and white, then it gets an uneven light gray to dirty beige shade, sometimes with flecks of yellow. Often cracked and irregularly bumpy with a wavy edge.

STIPE: 1–2 inches (3–6 cm) tall and centered, although sometimes off-center. Lighter than the cap but can have a darker tone down toward the narrower base.

PORES: Shallow with lots of small pore openings and a pore surface that is impossible to remove. Somewhat decurrent. First white, then shifts into a pale yellow.

FLESH: White to pale yellow, dense and chewy. Gets thick yellow patches when bruised.

SPORE PRINT: White.

RANGE AND HABITAT: Widespread in cool regions of North America. Found growing with spruce in the mountainous areas of the Western United States and in New England; even more common in Canada. Grows August to October, typically with spruce and often in large groups, crooked lines, or fairy rings.

SMELL AND TASTE: Tangy and mild.

PICKING AND CLEANING: As there are lots of Sheep Polypore in the same area, choose the youngest ones with the palest cap surfaces. Cut off the earthy stipe and divide the mushrooms to check that they are not infested.

PREPARING AND STORING: Mixes well with other mushrooms, especially with the tastier varieties, as the Sheep Polypore soaks up the flavors. A clear indication that you have a Sheep Polypore is that the flesh turns a yellow green hue when heated. It also turns very black when prepared in a cast iron pan. The flesh has a pleasant and firm consistency. If it's dried it needs to be cut into very thin slices.

LOOK-ALIKE MUSHROOMS: *Albatrellus confluens* is not an equal when it comes to being edible. The cap is a shade of apricot and doesn't crack with age but gets small pockmarks instead. It grows in dense groups.

Closely growing *Albatrellus confluens* can create a surface up to 20 inches (0.5 m) in diameter.

GIANT PUFFBALL

THE SUN IS slowly setting as Sweden's beautiful Sörmland landscape passes by the car window. Suddenly something catches our eye: an enormous, almost luminescent white egg. Our brakes screech to a halt and we pull a U-turn.

Lush grassy areas surround a Giant Puffball that stands there in all its pride, fully grown with an ivory coat puffed up in a cool sci-fi shape. The person who cut the grass here clearly knew that this beauty should be appreciated. Often these rare mushrooms live a dangerous life in parks and near built up areas, where they seldom get to grow this big because of feet and lawnmowers.

The surface feels like the underside of a sheepskin and in the cracks you can see that the spore making process is fully underway. This means that it has passed its sell-by date as an edible mushroom. A piece of string, that we later measure, shows that the fruiting body has a whopping 67 inch (145 cm) circumference.

To start with, the fruiting body grows by more than 20,000 cells per minute. In terms of the circumference, this means an increase in size of ½ inch (1 cm) an hour.

The World's Largest Ground-Growing Mushroom

Giant Puffballs in central Stockholm.

Fried Giant Puffball seasoned with some sea salt—what a delicacy!

WHEN THE POLICE were called to a south London work site where the workmen claimed to have found skulls under a floor, it was the experts from the Royal Botanical gardens who cleared up the mystery: they were Giant Puffballs.

There are many stories, as well as records, noting how large and heavy these fruiting bodies can get. In normal cases the Giant Puffball can grow as large as a football and weigh over five pounds, but it sometimes grows much larger and then can weigh more like forty pounds.

THE GIANT PUFFBALL can be found in across much of the world's temperate zones with a varied terrain, both in the northern and southern hemisphere. Several related species occur across much of North America. This mushroom prefers to grow in fairy rings in lime-rich soil and fertilized ground.

They are edible when young, when the flesh is white and firm, but you should remove the chewy outer surface before cooking.

The fruiting body grows quickly and creates trillions of spores that wait to float away, though in most cases it's only a single spore that manages to find a home with all the right conditions.

Giant Puffball *Langermannia gigantea*

FRUITING BODY: 8–24 inches (20–60 cm) or more wide. When fully grown can weigh up to 50 lbs (23 kg). The Swedish record was noted in 1997 on the island of Gotland with a weight of approximately 42 lbs (19 kg). Oval to round in shape or irregularly round. At first the surface looks like a prepared sheepskin but gradually small cracks appear, discolor, and finally turn brown and shriveled. With age the cap surface "blossoms" into irregular tabs that expose the mature spores.

FLESH: Nearly the whole inside consists of gleba where the spores are created. In the beginning the flesh is compact, white to yellow white, and then the spores mature and change color from yellow brown to an fine olive brown powder.

SPORES: Olive brown and spread through touch, for example by rain and wind. When the mushroom has dried it can release itself from its lower surface and roll around, whereby the spread of spores continues.

RANGE AND HABITAT: From mid-summer to mid-fall, often in scattered groups or fairy rings in fertilized soil in parks or pasture or along the edges of fields. Common in many regions and prefers areas with sweet limed soil.

SMELL AND TASTE: Mild and pleasant. When cooked, tastes like a nutty pancake or tofu.

PICKING AND CLEANING: Only pick the young, firm ones that have an all white flesh. Cut off the base. Don't use mushrooms that grow near traffic.

PREPARING AND STORING: Best to fry or deep fry in slices. Remove the chewy outer skin before cooking. Not suitable for drying.

LOOK-ALIKE MUSHROOMS: *Mycenastrum corium* is also edible and is a species occurring around livestock areas. It does not grow as big as the Giant Puffball and cracks into squares when fully mature. There are a number of other large and small Calvatia Puffballs and many are good edibles when collected young and firm.

A divided Giant Puffball where the production of spores has begun.

The fruiting body's sterile base is not spore producing.

The same example as on page 106–107. After two weeks the outer surface cracked.

WARTED PUFFBALL

PUFF, PUFF, PUFF! It's impossible not to step on old puffballs—in this case, the warted variety. It's great fun when the spores create their "smoke signals" and this is the whole point: for them to spread and find new places to grow.

Slightly younger examples can be found close by, creamy white and beautifully sprinkled with small conical spikes that are surrounded by even smaller warts. A few spikes are chocolate brown at the tip. We gently squeeze the cap and it feels soft. We divide the mushroom, and the flesh at the top has got a pale beige shade. It has started to produce spores, which means it's too old for our kitchen. More fruiting bodies appear from the carpet of needles and their flesh is white. We brush them off before the Warted Puffball is placed in our basket.

It's called the Gem-studded Puffball in North America; perhaps this is a more appropriate name?

◁ Young Warted Puffballs are like small, artistic jewels from the forest.

▯ The fine spikes and bumps eventually fall off and create an opening at the top where the mature spores can "puff" out.

A Smoking Mushroom with a Chimney

THE WARTED PUFFBALL is a popular edible mushroom that grows in both coniferous and deciduous woods. It's easy to recognize and is never infested, and it has no dangerous look-alike mushrooms, unless of course you were to mistake it for a very young Fly Agaric (see photo on the next page). When divided, you can see that the fruiting body of the Fly Agaric has the predisposition for both a cap and stipe, which the puffball does not.

Older fruiting bodies of the puffball are a beautiful bronze color before the spikes and warts fall off. The opening on top where the spore powder will be released has begun to form.

Here you can see the tiny warts between the conical spikes.

The top part shifts in shades of beige. Spore production has started and the bottom part, which is sterile, eventually dries to a woolly mass. In the olden days, this mass would be used as a compress to stop bleeding.

Warted Puffball *Lycoperdon perlatum*

FRUITING BODY: 2–3 inches (3–8 cm) tall. Pear-shaped and mostly covered in spikes and tiny skittle-shaped warts that fall off when touched. At first snow white, then yellow, and finally brown. The thicker tip part, with the spore producing gleba, rounds out with age. The transition to the sterile bottom part can be slightly wrinkled. The spokes and warts fall off the older mushrooms and when the spores are mature, a small hole opens at the top for them to "puff" out.

FLESH: First white and firm, then yellow and finally brown, which is when the powdery spores in the top and bottom part have dried into a woolly mass.

SPORE PRINT: Olive brown.

RANGE AND HABITAT: Widespread and common across North America from July to October in coniferous and deciduous forests and even in fields and on rotting wood. Usually grows in small groups.

SMELL AND TASTE: Mild and pleasant.

PICKING AND CLEANING: Only pick the young, firm mushrooms. Cut off the base of the stipe and brush off spikes and warts. Divide the mushroom and check that the flesh is completely white. If spore production has begun, it will taste bitter and is not suitable to eat.

PREPARING AND STORING: Works best with other mushrooms. The outer surface does not need to be removed.

LOOK-ALIKE MUSHROOMS: None. The Pear-shaped Puffball, *Lycoperdon pyriforme*, is bigger and has a grainy surface, and is equally valued as an edible species.

Long-stemmed Puffball.

The young fruiting bodies of the Fly Agaric have been confused with the Warted Puffball. When cut, you can see that it is a gilled mushroom with a stipe. The photo shows the beginning of a brown or red Fly Agaric.

PIG'S EAR

"DO YOU HAVE anything else to wear?" Our mushroom guide, Dieter Endom, warns us against the mosquitoes before we head off into the forest where we've been assured that we will find some Pig's Ears: the rare mushroom that few have seen, picked, or even tasted. And yes, it truly is an excellent edible mushroom.

The late July weather is perfect: sunny with fluffy clouds, and the tall spruce and a few leafy trees shield us from the beating sun.

The vegetation is rich in herbs and moss and is unique to this type of old pasture-land, which is lime rich and teeming with blood-thirsty mosquitoes.

Soon we see a long arch filled with densely growing violet-colored mushrooms.

They are so beautiful and we humbly fall to our knees in front of these rarities that are unlike any other mushroom.

"This mushroom is related to the Chanterelle and the taste is reminiscent of meat," Dieter explains while picking a bunch of mushrooms for our tasting session.

After a delicious meal, I still have to say that I think the Chanterelle tastes better and in any case, we don't want to overdo it when it comes to this rare mushroom.

◁ Lots of lovely Pig's Ears in a 5-½ yard-long (5 m) arch.

◻ We enjoyed a delicious meal with Pig's Ear as the main ingredient.

The spread of Pig's Ear in northern Europe is concentrated to the northeast coast of Uppland, in Sweden.

Beautiful and Vulnerable

THE PIG'S EAR only has a few thousand known locations in Europe where it grows and most of these are beech forests. In Sweden, the mushroom mainly grows in coniferous woods with lots of spruce and limey soil. These types of terrain are starting to disappear and they contain not only rare herbs but mushrooms that are close to extinction. They live in symbiosis with their host trees, which makes them vulnerable to modern forestry, and the Pig's Ear is a species that is now being actively preserved.

PIG'S EAR WAS previously known as the club trumpet mushroom in Sweden and was part of the *Craterellus* genus; now it's part of the genus *Gomphus*, and is the only one of its kind in northern Europe, though there are two other *Gomphus* in North America. It is loosely related to the Chanterelle, *Clavariadelphus pistillaris*, and *Ramaria*.

Melampyrum sylvaticum and Pig's Ear thrive in lime-rich soil.

Pig's Ear *Gomphus clavatus*

FRUITING BODY: 2–4 inches (5–10 cm) tall, meaty and reminiscent of a cornucopia shape.

CAP: 2–3 inches (5–8 cm) wide. Smooth, shallow, and cone shaped. Matte, with an irregular, wavy margin that often forks out. The color varies between dirty yellow, olive, and gray-purple. It has a clearer violet shade toward the edges. The pretty violet color is strongest on the younger fruiting bodies but as it ages, the color becomes more of a gray beige.

OUTSIDE: Shallow, fork-shaped, wrinkly ridges that turn smooth just inside the margin. Violet to flesh-colored pink.

FLESH: Marbled white and compact.

SPORE PRINT: Ocher yellow.

RANGE AND HABITAT: This unusual and infrequently found mushroom grows mainly in older coniferous woods across cooler regions of North America, from the West Coast of Northern California to New England and into Canada. Grows in large groups, often in a curve, from the middle of July to the end of September.

SMELL AND TASTE: Mild and pleasant. Avoid older mushrooms as they are bitter. Leave them to spread spores instead.

PICKING AND CLEANING: Is rarely infested. Cut a bit off the stipe as this can be bitter. It should not be picked in large amounts as it is so rare.

PREPARING AND STORING: The mushroom is good when fried, preferably in larger chunks. If the mushroom is to be dried, the pieces should be thin, as the flesh is very firm.

LOOK-ALIKE MUSHROOMS: The other species of *Gomphus* include the orange Scaly Vase, *Gomphus flocosus*, and the unusual tan and scaly *Gomphus kauffmanii*. Some people eat the Scaly Vase, though is is not prized. Neither have the unique appearance of the Pig's Ear.

The Pig's Ear turns more of a piggy pink color on the outside with age and the violet color completely disappears.

CLUSTERED CORAL

Our target is picked—our first Clustered Coral. It has long been on our wish list and it's a young one, which is perfect for our pan. The mushroom tastes mild and pleasant with a good, firm consistency.

WE'RE FEELING THE strain in our legs. We've walked in this harsh forest terrain for hours and we're thinking more and more about the hearty picnic that's waiting for us in the trunk as we start to wander toward the car.

"Come and have a look! A strange Ramaria."

It means turning back but I'm curious and a strange thing appears in front of me. A red, coral shaped creation with a stark white stipe is shining as strong as the sun. I'm struck speechless by this ethereal mushroom and it slowly dawns on me what this is—a species we have never encountered before.

"It's a Clustered Coral!"

Like Broccoli and Fennel

YOU SHOULD AVOID picking *Ramaria*, as they're hard to distinguish between, and those that are edible unfortunately have some dangerous look-alike mushrooms.

There is, however, a species that is safe to pick; this is the exquisite Clustered Coral. If you find a young mushroom, it's easy to recognize as the bushy branches are strongest in color at the tips and are pink/red. The fruiting body moves into a stark white, fleshy stipe.

You could liken the shape of the top to broccoli and the bottom to fennel. In time, the color fades and the mushroom becomes harder to recognize, and it's only the young specimens that taste good.

It has long been thought that the Clustered Coral is, in fact, part of a complex of several varieties that are bound to different types of trees.

Clustered Coral *Ramaria botrytis*

FRUITING BODY: 4–6 inches (10–15 cm) tall and often just as wide, almost globe shaped. It has a thick, white bottom part that goes into a thick forked network at the top, which in turn forks into fine, blunt ends that are a bright pinkish red. With age they get paler and yellow and after some time, the Clustered Coral looks more like other species of *Ramaria*.

FLESH: White, marbled and firm but still tender and often very infested by insects.

SPORE PRINT: Ochre yellow.

RANGE AND HABITAT: Found most commonly in the mountainous western United States growing with conifers and tanoak. Grows during the summer months. It can appear in large groups if the conditions are right and can even be found in mixed woods.

SMELL AND TASTE: Mild and pleasant.

PICKING AND CLEANING: Only pick the young mushrooms as the old ones are bitter and more easily mistaken for toxic species of *Ramaria*. Carefully cut around the base of the stipe to loosen the mushroom from the ground in order to get all the flesh. Peel off the earthy base and divide. Cut off any parts that are infested by maggots. Divide the fruiting body into smaller pieces and rinse them in running water if required.

PREPARING AND STORING: Works well when prepared most ways, even when deep fried. Bitter when dried.

LOOK-ALIKE MUSHROOMS: Use caution! None when young, as the tips are a bright red-pink color. Never pick any other types of *Ramaria* and only pick this mushroom when young and the bright pink-purple tips contrast with the white stipe!

Ramaria are still a quite unknown group of mushrooms. Not even the experts can differentiate between species of *Ramaria* with any certainty without a microscope.

WOOD CAULIFLOWER

The Wood Cauliflower's short, bowling pin-shaped stipe was firmly established in this tree stump.

An older Wood Cauliflower was found at the base of a living pine tree.

Wood Cauliflower
requires living or
dead pine trees
to survive. Here's
one by an old tree
stump.

I'M NOT QUITE sure what I'm seeing deep in the pine forest. A bit further along the terrain, which is covered in dry needles, I see a light ball by an old pine stump and I approach it with high hopes.

It's the middle of September and the forests are packed with mushrooms. My basket is filled with delicacies like chanterelles, boletes, and tooth fungi.

Now my focus is on a more unusual delicacy: the top prize in the mushroom lottery.

I part some dry spruce branches that are trying to grab at my sunglasses and remove the spider webs that stick to my face, to better focus on what I'm looking for.

Bingo! It's a Wood Cauliflower and it is stunning. The ornately decorated caps form a soft and round coral-like shape. This mushroom really is like no other and could probably inspire a clothing designer or two—maybe to create a new bathing cap!

YOU CAN TELL the fruiting body is young from its size and color. Older specimens turn yellow, then yellow brown, and taste bitter. Cauliflower mushrooms can grow really large and although this one is small and neat, it's still big enough to make a delicious meal for several people.

We need to be careful now; picking the whole fruiting body without damaging it can be tricky, as the flesh is tender and delicate.

I take the mushroom knife and carefully cut around the base in the ground beneath the mushroom ball. I can feel that the submerged stipe has grown tightly into the woody fibers of the tree stump and this is where I have to pull, tug, and jerk at it.

OOPS, THE STIPE just broke, but it needs to be rescued as the flesh in the stipe is great to eat. The rest of the cleaning job will be the most difficult, because the inside, which is very similar to a cauliflower, has irregular twists and pockets that most likely contain some dirt and insects.

I carefully lift up this rare mushroom with both hands. It smells of pine and turpentine and so the host tree has given its flavor to that which destroys it.

A Long-Lasting Mold Killer

IT'S CLEAR THAT the cauliflower-like
appearance of the fruiting body has given
this mushroom its name.

The Cauliflower mushroom is not very
common and is spread over larger parts
of the temperate zones in the northern
hemisphere. In Sweden, this variety grows
mainly with pine in the southern and central areas during the fall. Cauliflower mushrooms are excellent edible mushrooms and
are appreciated all over the world. Asia has
recently succeeded in cultivating it.

Even if the color, shape, and size vary, the
Wood Cauliflower is easily recognizable
and is unique in its appearance as the fruiting body looks like a natural sponge.

THIS SPECIES LIVES both as a parasite and
as a saprophyte on conifer trees, which
means it lives on both living and dead wood.
It prefers older forests where it lives at the
foot of the tree trunk or on tree stumps
where the mushroom eventually creates
blight in the wood.

Cauliflower mushrooms can often grow
very large and heavy. The latest record is
close to 66 pounds (30 kg) and this example
was found in France in 2000. The mushroom
has a distinct aroma and smells like a paint

Wood Cauliflower
contains a substance
that kills off mold
and, when cleaned,
keeps fresh for
several weeks in the
refrigerator.

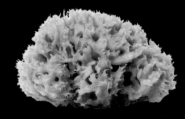

A Wood Cauliflower or a natural sponge?

A Cauliflower mushroom that has been divided and rinsed. The dense central section has absorbed water, creating the marbled effect on the flesh. The grit that is still attached has to be cut away.

shop or a new pair of leather shoes. The Cauliflower mushroom is very interesting within medical research; it has been discovered that the Cauliflower mushroom contains an antibiotic substance that kills mold, as well as containing substances that prevent cancer and stimulate the immune system.

THE CORRECT TERM is Wood Cauliflower, as there are several species in the genus *Sparassis*. One example is *Sparassis laminose*, which is very rare and grows in central and southern Europe. It was once found in Sweden, but is now extinct here. The species was mentioned in the celebrated Swedish lexicon, *Nordisk familjebok* (The Scandinavian Family's Book) from the 1800s and was then called the "straight" Cauliflower mushroom.

Wood Cauliflower *Sparassis crispa*

FRUITING BODY: Large and shaped like a head of cauliflower. 4–16 inches (10–40 cm) wide with wavy spirals that grow on cauliflower-shaped forks from a thick central column. At first creamy white, then yellow to ochre colored. Can weigh several pounds.

STIPE: Grows beneath the earth. Short, flattened, and club like. Also tastes good.

FLESH: Springy and waxy, tender in the outer parts and tougher in the center and stipe.

SPORE PRINT: White.

RANGE AND HABITAT: Not very common. It appears occasionally during summer and fall in older coniferous forests across the United States and Canada, usually at the base of pine or Douglas Fir trees. Grows on dead or living wood.

SMELL AND TASTE: A slight smell of turpentine and new leather. Wood Cauliflower fried in butter can resemble an omelet in taste, but it has a special taste that is considered delicious by some while others don't like it at all.

PICKING AND CLEANING: Try to get the whole mushroom up together with the stipe that is below the earth, as this also tastes great. Remove the bits covered in earth while you're still in the forest. Divide into larger pieces to start with and rinse in cold water. Then divide into smaller pieces that can be left in cold water for about 10 minutes, as this will make debris and insects float to the top. If you can, rescue the little critters and put them back into nature. Rinse the mushroom pieces again and remove earth and grit with a clean scrubbing brush. Finally, cut away any grit and other bits that are firmly attached. It's fiddly but well worth the effort.

PREPARING AND STORING: Fresh, young mushrooms taste best; the older ones are bitter. Fresh, cleaned Cauliflower mushrooms stay fresh in the refrigerator for at least two weeks. The flesh will keep most of its consistency when cooked and it is good when fried, stewed, or deep fried. When dry, it can be used to make mushroom flour.

LOOK-ALIKE MUSHROOMS: None.

The color of the flesh changes from creamy white to orange brown and the edges are often darkest at the ends.

BLACK MOREL

The Black Morel is one of spring's first edible mushrooms. The small fruiting bodies can be hard to see in the litter of last season's forest.

WHEN I ASKED my big sister if she'd seen any Black Morels on her lawn her answer was simply, "We ate them." Lucky for us, she didn't manage to find them all.

Spring arrives late in the archipelago and you can see that the birch trees have not yet blossomed. A pale May sun peeks through fast-moving clouds. Thin lily of the valley leaves, wood anemone, and budding blueberry bushes give us hope for the summer to come.

The Black Morel is one of the mushroom season's first delicacies. It's similar to the False Morel, its poisonous cousin, in color and consistency. However, the shape is different and it doesn't need to be parboiled, which the False Morel must be before eating to remove the poison.

What does it taste of then? Like the False Morel, only slightly milder.

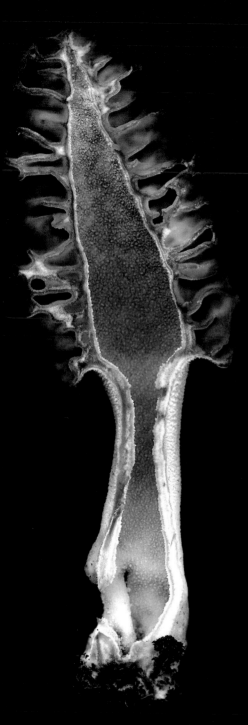

A Small Edible Mushroom with a Large Market

COLOR AND SHAPE within the morel family varies and it's still not clear how many species there are. The Black Morel grows pretty much all over the northern hemisphere and even in North Africa and Australia.

Black Morels are excellent edible mushrooms and are exported to several countries, including India and China.

THE MOST COMMON species in Sweden is the Black Morel. The cap, or rather the long "hood" as it resembles, is symmetrically shaped with a network of small holes between vertical edges that turn dark when the weather is dry.

The Black Morel is not poisonous like the dangerous False Morel, but if you are even the least bit unsure of what you have picked, parboil the mushroom.

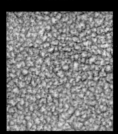

The Black Morel's stipe and the whole inside has a unique structure that is reminiscent of washed fleece.

A very rare species is the Bell Morel, *Verpa bohemica*. It grows with Aspen and other poplars and alders and is common in some regions of the United States and Canada.

Black Morel *Morchella conica*

FRUITING BODY: Small and thin and can be hard to see from above. Hollow. 4–6 inches (10–16 cm) tall, and ½–2½ inches (1½–6 cm) wide. The edge between the cap and the stipe grow into each other.

CAP OR HOOD: 2–5 inches (5–12 cm) tall and 1–2 inches (3–6 cm) wide. Medium to dark brown and sometimes gray to olive green in color. Pointy pyramid or egg shaped with vertical ridges that are separated by small, hollow chambers. The surface on the inside is creamy white and grainy.

STIPE: 2–4 inches (5–10 cm) tall and 1–1½ inches (2–4 cm) wide. Creamy white and grainy with a thicker base.

FLESH: White, thin, and tender.

SPORE PRINT: Pale yellow.

RANGE AND HABITAT: Grows in groups from April to the end of June in coniferous and deciduous forests. Also grows in open fields and among timber in shards of bark. Is fairly widespread.

SMELL AND TASTE: Mild. When cooked the taste resembles the delicate but poisonous False Morel.

PICKING AND CLEANING: Pick carefully as the flesh is very delicate and breaks easily. Divide the mushroom as it is often filled with snails, slugs, or sal bugs, more commonly known as pill bugs in north america.

PREPARING AND STORING: The best taste is achieved if you dry the mushroom.

It works well when stewed or in sauces. Black Morels do not contain any poison but as it can be confused with the False Morel, it is often recommended to parboil it if you are in the least bit concerned.

LOOK-ALIKE MUSHROOMS: The Common Morel, *Morchella esculenta*, fruits later than the Black Morel in most areas and is the most frequently eaten wild mushroom in the United States. The cap is yellow and rounder in shape. The False Morel, *Gyromitra esculenta*, is deadly poisonous. It has a more irregular cap shape and is larger in size (see page 210). The edible Half-free Morel, *Morchella semilbera*, is similar, but with the cap partially free from the stipe.

Common Morel.

False Morel.

WEEPING MILK CAP

"STOP! LOOK AT those terracotta colored mushrooms, I wonder what they can be?"

We move through the small forest, which is a well known spot to us that has yielded many mushroom harvests over the years. But these little beauties are new to us.

A page from my first mushroom guide slowly comes back to me. I have looked at the pictures and read the description many times, but never have I found this highly prized Weeping Milk Cap.

The color and shape matches, the fruiting bodies are chubby and irregular, and some of the specimens are of an impressive size. The caps look like burnt, unglazed clay in various shades of brick red. They are darker in the middle and toward the edges fine cracks appear. The thick silky stipes are lighter and more apricot in color.

I pick one and it is dense and heavy. Both the cap and stipe are velvety and dry. A snail, perfectly camouflaged with the color of the stipe, is feasting on it. I divide the fruiting body and place the pieces on the moss. Touching it causes an immediate effect and a white liquid seeps through in increasing amounts from the creamy white gills.

I put the mushroom through the last test: the liquid definitely smells of fish and shell fish. I carefully taste the liquid; it's not as sharp as the Rufous Milk cap and the cap is too brightly orange to be the Fenugreek Milk Cap. This clearly is a Weeping Milk Cap.

We find more places where we can fill our basket; it clearly is a great year for this rare species.

TIME TO CLEAN, prepare, and more importantly, taste the mushroom. The flesh is very hard—almost like cutting through an eraser. When you roll it in your fingers, the sticky liquid will harden into little balls.

The mushroom pieces fight back in the pan and take half an hour to start to soften.

My taste buds sharpen and my mouth is filled with the buttery crab-like smell that quickly turns to the scent of a bitter, dark chocolate.

The Weeping Milk Cap's milky liquid smells of cooked shellfish. Snails and slugs like to eat this mushroom, but it's rarely attacked by maggots.

A Beauty That Smells of Shellfish

THE WEEPING MILK Cap is fairly rare in Sweden. It grows in southern and central Sweden but barely exists north of the Swedish area of Mälaren. It grows alone or in small groups, some years in larger quantities. It's often found in older forests that have been used as pasture and that have not been logged for many years.

Their season is from the end of summer through to late fall. This species can also be found in other parts of Europe, eastern America, and Japan. In many areas, the Weeping Milk Cap is making a comeback, and this is thought to be due to environmental factors. In Holland and Denmark it's considered an endangered species. In parts of the United States it's a common and popular edible.

The Weeping Milk Cap is a mycorrhizal mushroom that chooses host trees with tough wood, such as the oak, beech, or hazel. It also thrives with spruce in lime-rich soil. Weeping Milk Caps are an indicator that the forest should be protected.

THE MUSHROOM PROBABLY got its name from the milky liquid it "weeps." In Sweden, it's known as the "Almond" Milk Cap due to its almond-colored fruiting body.

The Weeping Milk Cap releases a white, milky liquid when touched or damaged. The liquid eventually turns brown and sticky. It contains natural rubber, which is produced by several

The Weeping Milk Cap is one of the forest's most beautiful mushrooms. The swollen fruiting bodies can grow substantially in size and the velvety caps often have strange pockets and bulges.

The milky liquid eventually dries and turns brown.

plants such as the rubber tree and dandelion, and even a few
mushrooms that belong to the *Lactarius* genus.

ALL MUSHROOMS IN the genus *Lactarius* release a liquid when
damaged—some a clear juice, others may be yellow, orange, or even
blue. There are some species that are reminiscent of the Weeping
Milk Cap but none that have such a bright orange color, hard flesh,
or that have the specific shellfish aroma. Many find it wonderful
while others reject it because of its smell, even before picking it.

Older mushroom guides tell you that the Weeping Milk Cap can
be eaten raw, but this is no longer recommended. In its raw form, it
can contain unhealthy substances that disappear when heated.

Weeping Milk Cap *Lactarius volemus*

FRUITING BODY: Thick and robust.

CAP: 2–6 ½ inches (5–17 cm) wide. Dry and velvety with a powdery matte texture and never sticky. Orange to red brown and starts off slightly convex but often develops a depression in the center. The margin stays rolled under for some time. Older mushrooms often have cracks with a darker shade in the middle.

GILLS: Creamy white, sub-decurrent, which means they travel partially down the stem, and are densely packed. Turn brown when bruised. White drops, released in copious amounts, often hang under the gills.

STIPE: Thick and 1½–5 inches (4–12 cm) tall. Tapers off and darkens towards the base. Matte with a light cap colored shade.

FLESH: White to pale yellow. Hard and rubber-like and turns brown when bruised.

LIQUID: Very runny when cut or damaged. White at first, then gray brown and sticky when dried. Smells of herring or cooked shellfish. Stains fabric and hands brown.

SPORE PRINT: White.

RANGE AND HABITAT: Mid-summer to fall, growing alone or in small groups in deciduous, mixed, and coniferous woods that are rich in herbs. Found especially with oak, and common in many Central and Eastern US forests.

SMELL AND TASTE: Has a distinct smell of fish and shellfish. Mild with a slightly bitter aftertaste. When prepared, you get a more muted taste of cooked crab.

PICKING AND CLEANING: Rarely attacked by maggots but often by snails and slugs. Thoroughly clean the mushroom at home to avoid the sticky liquid in the forest.

PREPARING AND STORING: Best to fry fresh for at least 20 minutes. Can be mixed with other mild tasting mushrooms. Can be frozen after having been parboiled or dried. Soak for 2–3 hours.

LOOK-ALIKE MUSHROOMS: Closely resembles edible *Lactarius corrugis*, but without the dark brown corrugated cap. Also closely resembles the edible and desirable *Lactarius hygrophoroides* with broader gills and a mild scent. Fenugreek Milk cap, *Lactarius helvus*. Mildly poisonous and dusky pink in color. Smells like licorice or curry with a clear juice. Rufous Milk Cap, *Lactarius rufus*. The mild poison disappears when boiled and it can then be consumed as an edible mushroom. The cap often has a hump in the middle and contains a white liquid.

The smaller Rufous Milk Cap is a dark, meaty red.

SAFFRON MILK CAP

◁ A whole family of Saffron Milk Caps

▢ An older example where the fresh orange tone has started to turn green.

THE SEPTEMBER SUN *shines across the pleasant woodland path and the weather is just right. The old pine forest is rich in lime, which is made clear by the ground's herby vegetation.*

A few bright mushrooms rear their heads among the wood sorrel and wild strawberry leaves. The caps, with their sunken middles, are patterned with concentric circles in pale pink and carrot red, and the largest ones have green patches.

I divide one and get an immediate reaction. Orange liquid seeps out of every part of the mushroom, and on the carrot colored stipe you can see small, shallow pockets in a slightly darker shade. Everything points to one thing, the delicious Saffron Milk Cap has been revealed.

Lactarius quieticolor (from Europe) has, despite its name, an orange milky liquid, but the cap can shift in shades of steel gray to blue.

Many New Species of Deliciosi Have Been Discovered

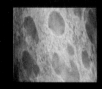

The stipe has a dimpled pattern.

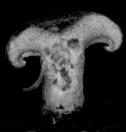

The porous flesh is often infested.

IT WASN'T UNTIL the end of the 1960s that it was discovered that a group of mushrooms known as Deliciosi in Scandinavia actually consisted of several species: the Saffron Milk Cap, *Lactarius deliciosus*, and the False Saffron Milk Cap, *Lactarius deterrimus*. Both are popular edible mushrooms but the Saffron Milk Cap is considered the better tasting.

Apart from these, there are two very rare species that only really exist on two islands off the coast of Sweden. These are *Lactarius sanguifluus* and *Lactarius semisanguifluus*.

Recent studies have shown that the Deliciosi in Scandinavia contains at least two further species: *Lactarius quieticolor* and *Lactarius fennoscandicus*. As of yet, there is very little information about the latter. Regardless of species, all Deliciosi are edible and their common denominator is that they all have an orange to wine-red milky liquid that either stays the same or darkens.

Saffron Milk Cap *Lactarius deliciosus*

FRUITING BODY: Smaller than the Weeping Milk Cap.

CAP: 2½–6 inches (6–15 cm) wide with a depressed center. First round and bumpy, then convex, and when fully mature is flat. The margin stays rolled in for a long time. It has concentrically zoned thin rings in orange and a somewhat wider zone in salmon to gray pink. Often has carrot colored patches and is dimpled toward the margins. Sticky or slippery when wet. When mature, it's often a coppery green color, especially after a cold period.

GILLS: Dense and sub-decurrent before abruptly ending. From apricot to carrot colored and slowly turn pistachio green when damaged.

STIPE: 2–3 inches (5–7 cm) tall. Thick and hard but also fragile. Is at first filled with a porous "marrow" in the center and then becomes hollow or has chambers. Has the same color as the cap and is patterned with different-sized shallow holes in a darker shade.

FLESH: Firm and carrot-colored. Releases a lot of liquid when damaged, which turns dark after an hour or so.

MILK LIQUID: At first brightly carrot colored, then turns a coppery green color after several hours.

SPORE PRINT: Light ochre.

RANGE AND HABITAT: The Deliciosi Group is fairly common across much of North America in areas rich in soil, mainly in coniferous forests. It is less common in other areas. Grows from August to October in moist woods and along waterways.

SMELL AND TASTE: Pleasant. Tastes very good when cooked. If you eat too much of it your urine might turn red, but this isn't dangerous.

PICKING AND CLEANING: It is almost always infested, especially the stipe, but you can salvage the better parts. With time they become so discolored that they look unappetizing, but the quality is the same.

PREPARING AND STORING: It tastes wonderful when fried fresh. They are not suited for drying, as the taste turns bitter. Freezing and preserving works well.

LOOK-ALIKE MUSHROOMS: There is a small number of similar orangeish *Lactarius* with orange latex that are edible. False Saffron Milk Cap, *Lactarius deterrimus* (page 147). Bearded Milk Cap, *Lactarius torminosus*, has a white, very acrid liquid and is edible only when well-cooked.

The Bearded Milk Cap has a paler color and a "felted" cap.

FALSE SAFFRON
MILK CAP

◁ The night frost has left these perfect False Saffron Milk Caps unbothered, as they are still protected by moss.

▯ The caps change color after a frosty night, from orange to green.

PLANTED SPRUCE LINE up in the snow. They're still quite small and won't be cut down for a while, and this also gives us a clue to the type of Deliciosi that appear here and there in the forest. It's the False Saffron Milk Cap, which likes the moist carpet of spruce needles. The False Saffron is an easy prey, but it's not quite so easy to get there before the insects do.

The bright orange liquid seeps through from the edges after the mushroom has been divided. After it has dried, it turns a wine red color. A few mushrooms at the edge of the forest have been affected by the frost and the caps are now almost completely green.

Many examples have to be discarded so it's just as well to sort them out here in the woods.

Once home, hot mushroom sandwiches with a chili-flavor await us.

A Popular Edible Mushroom with a Hint of Chili

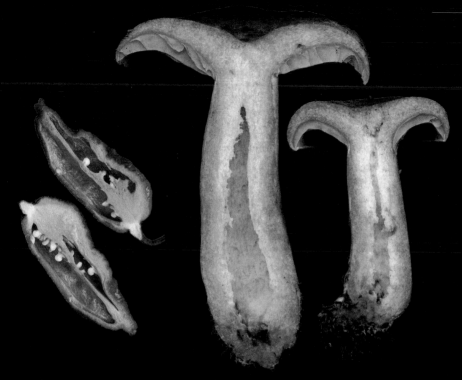

When the False Saffron Milk Cap dries, the flesh deep inside turns a lighter color and the orange liquid shifts to a dark red.

SOMETIMES WHEN YOU see them individually, it can be hard to distinguish between the Saffron Milk Cap and the False Saffron Milk Cap. They have different host trees, which gives you a clue to which species you're looking at. The Saffron Milk Cap is considered better tasting than the False Saffron Milk Cap. But some think the latter is an excellent edible mushroom, and others don't like it at all. Maybe it's the chili taste that burns slightly that puts people off?

The harsh and spicy flavor contains a varying degree of bitterness with a hint of sweet umami.

DIETER ENDOM, CHEF AND MUSHROOM EXPERT

False Saffron Milk Cap *Lactarius deterrimus*

FRUITING BODY: Like the Saffron Milk Cap.

CAP: 3–6 inches (7–15 cm) wide. At first round with a slightly depressed center and a rolled-in margin. Then shallow to severely funnel shaped. Sticky when damp. Has orange colored zones that turn greener with age and zones or patches that often go from the center and outward. The frost changes the color to green.

GILLS: Dense and descend slightly down the stipe. Same color as the cap.

STIPE: 4–6 inches (10–15 cm) tall. Hollow or with a loose flesh that is often infested by maggots. The same color as the cap and smooth, usually without patches or pock-marks. Tough, but at the same time delicate and fragile. Thinner toward the base.

FLESH: A dirty gray color. At first turns orange at the edges because of the milky liquid that escapes when divided or bro-ken, then darkens into a blood or wine red. Thinner than the Saffron Milk Cap.

LIQUID: At first orange. After 10–15 minutes turns wine red and finally a tarnished green color.

SPORE PRINT: White to light ochre.

RANGE AND HABITAT: Grows widespread across North America in association with spruce and often grows in vast quantities in damp soil. Prefers areas with lots of planted spruce and mossy forests and is found from August to October.

SMELL AND TASTE: Spicy and pleasant. When cooked, it tastes peppery. If you eat too much of it, your urine could turn a red color but this is completely harmless.

PICKING AND CLEANING: See the Saffron Milk Cap on page 143.

PREPARING AND STORING: Adds some spice to sauces, soups, and so on. Bitter when dried.

LOOK-ALIKE MUSHROOMS: Saffron Milk Cap, *Lactarius deliciosus* (page 143). Bearded Milk Cap, *Lactarius torminosus*, has a white milky liquid and is edible after it's been boiled (page 145).

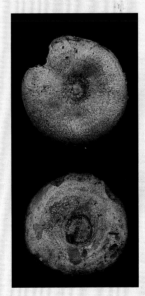

The cap at the top belongs to a False Saffron Milk Cap. It's often a coppery green color in the middle and is not as clearly zoned as the Saffron Milk Cap, in which the cap is usually more orange-pink with wider zones underneath.

BARE-TOOTHED RUSSULA

◁ Two fresh
Bare-toothed
Russulas that
have so far
survived any
infestation.

▯ An old, dry
example.
The top part
of the gills
can be seen
where the
skin of the
cap does not
cover the
edges.

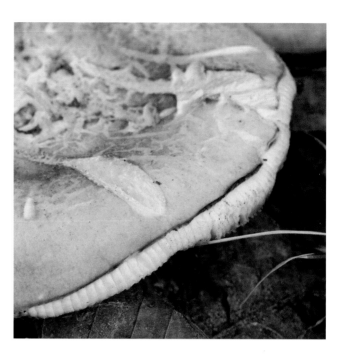

THE LATE SUMMER *rain keeps pouring, which has made for some densely-growing mushrooms in the woods. Many caramel colored caps can be seen and there is no doubt that these are Russula, but past this we are lost among the range of colors around us. They go from dark brown, wine red, red brown, golden red, olive green, pink, ochre yellow, yellow... the list goes on.*

However, there are some species that are slightly easier to detect—for example the Bare-toothed Russula. Apart from the distinctive pork-pink color, the Bare-toothed Russula also has the well-known characteristic of the skin of its cap not quite reaching to the cap margin.

I try a small piece and spit it out. It tastes neither sharp nor burning, so I'm one hundred percent sure of this one.

A culinary delight awaits us; butter fried Bare-toothed Russula tastes of nuts and bread.

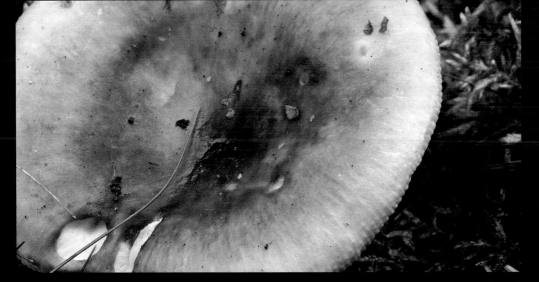

The Bare-toothed Russula's cap shifts in color from light to dark pink. When damp, earth and other rubbish will stick to it.

Wearing a Cap a Size Too Small

The stipe is usually shorter than the width of the cap and is hard like a wine cork.

WITHIN THE MUSHROOM flora there is an especially complicated genus called *Russula*. *Russulas* are hard to categorize if you're not an expert; on some species, cap colors can change dramatically after a downpour and many species are very similar in appearance.

However, many *Russulas* make very good edible mushrooms, and no species is dangerous. It's worth trying to learn a few of the approximately 150 different types that exist in the northern hemisphere.

THE DELICIOUS BARE-toothed Russula is a great mushroom for beginners. It's not as bright red in color as some of the sharper-tasting *Russulas*. Its colors shifts from plaster pink to flesh pink and its most distinguishing feature is the cap's skin, which doesn't quite reach the margin and reveals the top parts of the lamella at the outer edge. These look like leering teeth.

Bare-toothed Russula *Russula vesca*

FRUITING BODY: Short with a wide cap.

CAP: 2–4 inches (6–10 cm) wide. Convex to flat. From light pink (cooked ham) to dark pink (pork) in color. Sticky in damp weather, and this is when dirt will attach to it. Usually, although not always, the skin of the cap does not reach the margin.

GILLS: White, then sometimes get brown patches. Clear drops sometimes hang off the edges of the gills in damp weather.

STIPE: 1¹/₂–2¹/₂ inches (4–6 cm) tall. White and tough. Can get brown patches. Slimmer at the base.

FLESH: White and dense, sometimes gets brown patches when old.

SPORE PRINT: White.

RANGE AND HABITAT: Found occasionally to frequently from July through to October. Prefers beech and oak trees and even pine forests and grassy field edges.

SMELL AND TASTE: Mild with a somewhat sweet smell. When cooked it is reminiscent of nuts and bread.

PICKING AND CLEANING: Everything can be used apart from the earthy stipe. When necessary, rinse under running water.

PREPARING AND COOKING: Works well mixed with other mushrooms and keeps its delicate consistency, even when cooked. When dried, it needs to be soaked for a long time.

LOOK-ALIKE MUSHROOMS: There are a number of other *Russulas* that have a similar cap color, but the Bare-toothed Russula has some very unique features that are hard to mistake. Very closely related to the *Russula cyanoxantha* group of edible mushrooms. If you are sure that you have a *Russula* but are not sure which species, you can always perform a taste test (see below).

DISTINGUISHING FEATURES FOR ALL *RUSSULA*
- It is a gilled mushroom, often with a colorful cap in one or more colors.
- The stipe is white with shifts of red. White to creamy colored gills.
- The flesh is white and tender. Not fibrous and has no liquid.
- Has no ring, volva, or spider veins.
  If all of the above fits, you can taste a small piece of the gills. Chew and then spit it out, but never do so within reach of children. If the taste is mild, then the *Russula* is edible and tasty. A sharp, burning taste means that it's not suitable as an edible mushroom.

COPPER BRITTLEGILL

The red Copper Brittlegill is easy to find in the green moss.

MANY SPECIES OF mushrooms have been cleverly camouflaged to hide in their native environments. But when it comes to the Copper Brittlegill, the complete opposite is the case. It's definitely not a mushroom that is hard to find. Its yellow/red cap stands out like a beacon, and perfectly complements the green vegetation that surrounds it.

The round, undeveloped caps look like colorful eggs sitting in the soft moss, just waiting to be picked. They grow in vast numbers, although there's no point picking the larger ones, as they're almost always infested by maggots.

The clean mushrooms in the basket are starting to turn gray. This is a sure sign that we're holding the Copper Brittlegill, which is comforting to us, as we're only just getting to know this multicolored genus.

A *Russula* That's Good for Beginners

The skin on the cap is often so thin that it reveals the contours of the lamella.

THE FLESH TURNS gray, just like it does on the Yellow Swamp Brittlegill (see following chapter); this doesn't look very appetizing, but the taste remains unaffected. The Copper Brittlegill is not considered as tasty as its yellow cousin and when cooked the flavor can be somewhat acrid. Nevertheless, it's a popular edible mushroom that's beloved by many. This may in part be because the Copper Brittlegill is one of the easiest species to recognize within this genus.

The Copper Brittlegill grows with pine and spruce in the northern hemisphere and also with birch trees in mountainous regions.

Copper Brittlegill *Russula decolorans*

FRUITING BODY: Average size within this genus.

CAP: 2–5 inches (5–12 cm) wide. When young almost completely round, then convex to flat. Often has a depressed center. At first bright orange-yellow and then a more subdued brick red, sometimes with a darker center and can even have shades of apricot. In contrast to many other *Russula*, it retains a fairly regular shape. Sticky in damp weather, and has a smooth skin that can be removed.

GILLS: At first white, then creamy yellow and finally gray.

STIPE: 2¹/₂–3¹/₂ inches (6–9 cm) tall. White and then gray. The flesh has a porous center.

FLESH: White and tender, when damaged or old it will turn gray.

SPORE PRINT: Creamy yellow to pale ochre.

RANGE AND HABITAT: Found in coniferous forests across cooler regions of North America in lean soil, with lichen and moss. Also thrives in rocky woodland. Grows from July to October.

SMELL AND TASTE: Mild and discreet, the younger specimens can be slightly sharp. When cooked, it tastes good and can sometimes be a bit bitter.

PICKING AND CLEANING: Everything can be eaten except the earthy stipe, which should be removed. Mature specimens are usually too infested. If the cap is very dirty, the skin can be removed.

PREPARING AND STORING: Mixes well with other mushrooms, fried or stewed. Can be parboiled and frozen. Not suitable for drying as it may end up tasting bitter.

LOOK-ALIKE MUSHROOMS: There are other types of *Russula* that have a similarly colored cap, but if you should pick the wrong one, it will still be a good, edible mushroom. Just make sure that the taste is mild (see text box on page 159).

Like most other mushrooms within the *Russula* genus, even the Copper Brittlegill can quickly become infested.

YELLOW SWAMP BRITTLEGILL

THE PATH WINDS its way over green pastures and a woody area—a spread of nature. In June, the Heath Spotted Orchid and the Lesser Butterfly Orchid will appear in this field, but now is the time for other non-protected and edible species to appear.

In the overgrown birch forest, several butter-yellow mushroom caps appear. The thin skin on the cap is almost translucent toward the margins where the contour of the gills can be seen.

The taste is mild so it can't be the Ochre Brittlegill. I scrape the stipe and after a while the flesh turns gray.

All the distinguishing characters have been checked off; this is the Yellow Swamp Brittlegill.

SOMETHING ELSE CATCHES our eye and we are perplexed as to what these green mushrooms might be. Have spores dropped down from another planet to create these monstrous fruiting bodies?

The explanation is, of course, much simpler. It's the parasite Hypomyces luteovirens, which attacks Russulas and sometimes deforms the mushrooms so badly that they become unrecognizable.

Hypomyces luteovirens is a parasite that attacks *Russula*, rendering them inedible. In this instance, the fruiting bodies have been so affected that you can't see what species they actually are.

Yellow as a Buttercup

A lot of debris can get caught on the caps, which can be hard to remove. This is easily remedied by simply peeling of the skin.

IN SWEDEN, THE Yellow Swamp Brittlegill was originally known as the "Mild Yellow Swamp Brittlegill" in contrast to the sharp-tasting species, which is now known as the Ochre Brittlegill (or, in Sweden, the Mustard Brittlegill). It tastes unpleasant and it should not be eaten. In older mushroom literature, it's considered quite tasty after boiling, but this is no longer recommended. The Yellow Swamp Brittlegill, on the other hand, is considered one of the tastiest *Russula* mushrooms .

The species can be found in northern Europe, North America, and parts of Russia.

Yellow Swamp Brittlegill *Russula claroflava*

FRUITING BODY: Average size among the *Russula*.

CAP: 2–5 inches (5–12 cm) wide. Bright yellow in color, which diminishes with age, and slightly sticky when damp. Thin, smooth skin that can easily be peeled off. Often has a depressed center.

GILLS: Notched. At first white, then creamy yellow and sometimes with patches from insect attacks. Older ones can have gray or black edges.

STIPE: 2½–3½ inches (6–9 cm) tall. White, turns gray with age.

FLESH: White and turns gray when damaged.

SPORE PRINT: Creamy yellow.

RANGE AND HABITAT: Widely distributed across forested regions of North America, in deciduous and mixed forests. Prefers damp areas and grows mainly with birch and sometimes Aspen or alder trees between July and October.

SMELL AND TASTE: Mild and pleasant. When cooked, nutty or almond-like in flavor.

PICKING AND CLEANING: You can use everything except the muddy stipe, which should be removed. Mature mushrooms are usually too infested. If the cap is too covered in debris, you should remove the skin.

PREPARING AND STORING: Tastes good both on its own and mixed with other mushrooms. Is tender and dries well.

LOOK-ALIKE MUSHROOMS: *Russula lutea* is smaller but is also a good edible mushroom. The Ochre Brittlegill, *Russula ochroleuca*, is a dirty yellow and is not suitable for eating, (see text box on page 159).

Russula lutea has sparse gills and the cap is only 1–2½ inches (3–6 cm) wide, usually with a stronger yellow color in the center.

RUSSULA ROMELLII

A multicolored *Russula* in its various stages. It thrives in coniferous woods.

THIS LOVELY BEECH *forest always provides us with interesting mushrooms. Today, we are surrounded by Russula of various stages of development, many of which we have never seen before. Despite the fact that they are all very different inside, they all seem to be of the same species.*

Convex, underdeveloped caps look like old potatoes, both in color and shape, and the larger ones have a mixture of colors in lilac red and yellow green.

This season our ambition is to get to know more Russulas, but this fantastic species seems to have escaped most mushroom authors. We turn to the Internet and a mushroom forum gives us our answer: it is a Russula Romellii (Multicolored Russula in Swedish). It's a delicate edible mushroom and is common in beech forests but less so in the mushroom guidebooks!

You can see the little potato-shaped cap behind the older, darker one in the foreground.

The flesh is yellower under the skin of the cap. Note the pink hue at the margins of the larger mushroom.

Colorful but Well Hidden

MUSHROOM GUIDES DESCRIBE many good, edible mushrooms within the *Russula* genus, but there's surprisingly little written about *Russula romellii,* despite it being an excellent edible mushroom. It's most frequently found in the southern and central parts of Sweden in deciduous woods that are rich in lime, and grows with oak or beech trees. It is not known how common it is, but it is more common on the continent where it thrives at higher altitudes.

When you take a look online, opinion seems to be divided as to its culinary worth. In Germany it scores highly but Italians find it quite uninteresting as an edible mushroom.

Russula romellii

FRUITING BODY: Large within this genus.

CAP: 2¹/₂–6 inches (6–15 cm) wide and fleshy. At first round, then convex, and finally flat with a depressed center. Often patchy and sticky when damp and usually zoned in many colors. Creamy white or yellow green from the center, and wine red and purple toward the margins. Fleshy pink, ochre, red brown, and green to olive green can also appear.

GILLS: Vanilla yellow and densely packed.

STIPE: 2¹/₂–4 inches (6–10 cm) tall. White, sometimes with shifts of pink or red, and thicker toward the base of the stipe.

FLESH: Creamy yellow under the cap skin. Creamy white and firm and more porous in the stipe.

SPORE COLOR: Ochre yellow.

RANGE AND HABITAT: This mushroom has been reported as rare in North America but can be found in deciduous forests, mainly with beech and oak. Grows between August and September.

SMELL AND TASTE: Mild and somewhat fruity or like fresh grass. Nutty when cooked.

PICKING AND CLEANING: You can use all of the mushroom except the earthy stipe, which should be removed. Divide the mushroom in the forest; if there is debris stuck to the cap you can rinse it under running water.

PREPARING AND STORING: Great to fry alone or with other mushroom varieties. The firm flesh retains much of its consistency after it has been cooked and it also dries well.

LOOK-ALIKE MUSHROOMS: Other *Russulas* can resemble *Russula romellii*. One example is the Charcoal Burner, *Russula cyanoxantha*, which is also a good edible mushroom with a chewy, rubbery lamella. (See text box on page 153).

Russula romellii has a regal color scheme, from gold to purple.

ST. GEORGE'S MUSHROOM

OUR EYES HAVE devoured the mushroom guide's description of the St. George's Mushroom, and now we're eager to find one. Apparently, it's a fantastic edible mushroom.

Here, in a park just outside Stockholm, the conditions are perfect.

The St. George's Mushroom wants soil rich in nutrients and lime and now, when spring has sprung, this mushroom has its season.

The May sun is warm, but the ground still has a chill. We walk around the green grass where dots of yellow dandelion and heath speedwell brighten the terrain.

UP ON A hill we can see dark curves on the grass. This looks promising, as we know the St. George's Mushroom grows in fairy rings.

We hurry to it and adjust our eyes; at the edge of the line of darker grass, we see lots of bumpy mushroom caps—big and small like perfectly baked meringues that are creamy white with shades of yellow brown. The cap shapes are chubby and irregular because they grow so densely. On the smaller specimens, the caps have rolled-in margins; they look like small button mushrooms and the consistency is firm.

Is this really a St. George's Mushroom? I smell the gills and the scent is right, like flour—yes, damp flour or maybe fresh, wet cotton.

This mushroom thrives in nutrient rich soil and the low, densely growing fruiting bodies usually create large fairy rings or arches.

I'll double check the mushroom guides again; after all, there are some dangerous look-alike mushrooms, such as the Livid Agaric.

THE FAIRY RINGS yield a good harvest. The flesh on the stipe of the St. George's Mushroom is threadlike and tender while the cap is more firm. Though many of the mushrooms have been infested by maggots, we still make many trips between the car and our treasure trove to offload our cache.

Other Sunday strollers glance over at us.

"Are you picking nettles?" an older man asks.

"No, St. George's Mushrooms."

Depending on the weather and wind, St. George's Mushrooms change color from creamy white to patchy yellow brown, and the skin of the cap often cracks in dry weather.

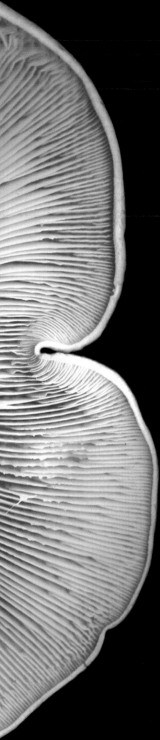

Can Be Harvested Many Times

THE ST. GEORGE'S Mushroom is not very widespread, and its main season is during May and June. It's fairly common on the Swedish islands of Öland and Gotland and exists all over Europe; they've even been found almost 8000 feet up in the Alps. It can also be found in North Africa. It does not occur in North America.

The species is usually found in groups in lime-rich lawns in arches or fairy rings. They can often be seen at a distance where the grass is a darker green.

Maggots tend to attack the fruiting bodies early on in the season—especially when the weather is dry and warm—but the mushrooms appear in large numbers, so the yield is still plentiful and it might be worth returning after a week has passed.

Older mushroom guides consider the St. George's Mushroom a delicacy compared to the Button Mushroom or the Porcini. Lately, however, the St. George's Mushroom has been pushed to the fringes of the mushrooming map and opinion is divided as to its taste. The flavor is very special and distinctive in dishes, which means that it's best when mixed with other, milder mushrooms that appear later in the season. While you wait for these, you can freeze the St. George's Mushroom; do so either fresh or after parboiling.

The St. George's Mushroom's fruiting body can be either thick or thin.

St. George's Mushroom *Calocybe gambosa*

FRUITING BODY: Usually thick and robust but can also be slim.

CAP: To start, white and ovate with a rolled-in margin. Convex when fully mature and irregular to flat with a folded out margin. Thick and fleshy and approximately 2–4 inches (5–10 cm) wide. It's often cracked with brown patches when the weather is dry but with age turns an even chamois color. Can sometimes be slightly pink as well.

GILLS: White, densely packed, and slightly notched.

STIPE: 1–3 inches (3–8 cm) tall. Short and chubby and can vary in width. Usually thicker at the base.

FLESH: White. The flesh of the stipe is tender and threadlike while the flesh of the cap is more compact.

SPORE PRINT: White.

RANGE AND HABITAT: Does not occur in North America. Not very common in Sweden. Grows during spring and early summer and sometimes as late as fall. It thrives in plush lawns, pastures, gardens, and so on. Usually appears in large groups in fairy rings or curves.

SMELL AND TASTE: A strong smell and taste of flour or wet cotton. When cooked it is reminiscent of the Button Mushroom or chicken stock.

PICKING AND CLEANING: Pick young, fresh mushrooms and remove bits that are infested. The whole mushroom can be used except the earthy stipe.

PREPARING AND STORING: Works well when fried and stewed and is excellent in soups. Can be dried in thin slices or parboiled and frozen. Can also be frozen fresh.

LOOK-ALIKE MUSHROOMS: The poisonous Livid Agaric, *Entoloma sinuatum*, (page 219). The cap often has a hump and it smells like flour, just like the St. George's Mushroom.

The Livid Agaric has a pink spore print (page 2).

ARCHED WOOD WAX

Some animal has chewed the edge of the cap to reveal the sparse gills.

◁ Younger and tougher examples that have the typical sooty brown cap.

A MAGICAL FOREST surrounds us. Rocks, tree stumps, and old, toppled tree trunks are covered with soft moss, and blueberry and lingonberry bushes can be seen scattered here and there. The spruces spread their healthy branches all the way to the bottom of their trunks and even the air has a green shimmer.

Our baskets are filled with Yellow Chanterelles and Yellow Foot and hours of cleaning are ahead of us.

Still we stop and examine some dark brown mushrooms. Thin, black rays streak from the pointy center toward the edges. The underside has sparse, white gills that travel down the streaky gray-brown stipe. This is the delicate Arched Wood Wax and as we look up, more fruiting bodies appear.

I sacrifice my jacket to serve a new function . . . as a basket!

The Arched Wood Wax prefers to grow in larger groups. The cap color varies from warm chocolate brown to brown black.

Dark Beauties That Thrive in Blueberry Woods

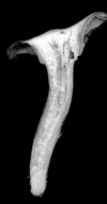

The Arched Wood Wax's flesh has a silky shine and is fibrous.

MOST WOOD WAXES come late in fall and this is true of the Arched Wood Wax as well. It's relatively common and grows alongside pine and spruce in nutrient poor soil between August and October.

It has a sweet, honey-like smell; the mushroom is well liked by animals, both four footed ones, as well as snails and insects.

Older mushroom guides didn't give the Arched Wood Wax the attention it deserves, but it was reviewed in the mid-twentieth century and today it's considered to be a delicious mushroom—one of the best in its genus. It's also larger in size compared to other edible wood waxes.

This species can be hard to detect, but it has few look-alike mushrooms and none of them are dangerous. The Arched Wood Wax is pretty easy to recognize.

Arched Wood Wax *Hygrophorus camarophyllus*

FRUITING BODY: When mature, can be skinny or fat.

CAP: 2–5 inches (5–12 cm) wide, convex to start with, with a visible hump in the center. The margin stays rolled in for a long time before spreading out and then usually becomes funnel-shaped. Chocolate or gray to brown-black in color. Black lines radiate from the darker center toward the edges. Lighter in color when mature and damp in wet weather, but it is never sticky.

GILLS: Thick, sparse, and decurrent. White at first and then gray to gray beige.

STIPE: 2½–5 inches (6–12 cm) tall. A similar color to the cap but with a lighter tone and threadlike texture. Grayer towards the base of the stipe.

FLESH: White to light gray and tender.

SPORE PRINT: White.

RANGE AND HABITAT: Found in cooler coniferous forests in both West and East coast states. Follows a similar distribution to spruce and likes nutrient poor soil in mossy coniferous woods. Likes rocky and hilly terrain and usually grows in larger groups. Typically fruiting between September and October.

SMELL AND TASTE: Sweet and pleasant. When cooked it has a pleasant, mild taste, but the flavor feel almost timid!

PICKING AND CLEANING: Unfortunately, maggots will quickly attack this mushroom. Later on in the season the fruiting bodies will fare a bit better.

PREPARING AND STORING: Good to mix with other mushrooms. When prepared on its own, it will need extra seasoning. It can be dried, frozen, or parboiled.

LOOK-ALIKE MUSHROOMS: *Hygrophorus atramentosus* is edible but rare and should therefore not be picked. The fruiting body is usually somewhat smaller and it has no brown flecks. There are several other dark *Hygrophorus* species in North America; all are considered edible, though not all are valued as edibles.

Hygrophorus atramentosus is a rarity in herbaceous spruce forests.

HERALD OF WINTER

The cap color varies in shade from olive green to a rusty brown and sometimes gray brown.

◁ On more mature specimens,
the sparse gills are a shiny orange.

THERE'S A HUNT going on! A yellow warning sign with a moose silhouette on it greets us on our path, but we have a completely different target in mind: a mushroom that appears in late fall that we have been promised can be found here. It's been on our wish list for a while: the Herald of Winter.

We've been disappointed so far, but we won't give up and so we keep moving toward some pine woods with flat rocky areas. We know the Herald of Winter thrives among lichen and moss.

The temperature fluctuates and the wind has a chill to it. The rain is tinged with snow and our clothes get wetter, our fingers colder, and our enthusiasm starts to wane.

The mushrooms finally appear. The caps are glassy with a darker center and the damp makes them extra shiny and sticky. They vary in color from rusty brown to olive green, even the gills. Some are orange and others pale yellow. For some strange reason, several of the fruiting bodies have a bright red patch.

A pleasant surprise awaits us when we get back into the warmth of our home: Herald of Winter fried in butter tastes a bit like chanterelles and a hint of lemon.

Picking Mushrooms in December

WHEN BOLETES, RUSSULAS, and other great tasting edible mushrooms have given up for the season, there are still some hardy species that can be harvested before winter takes a serious hold.

One of them is the Herald of Winter. It actually needs a few frosty nights in order to start creating fruiting bodies and it's usually the seasoned mushroom pickers who most appreciate this late mushroom.

It grows all over Scandinavia on sandy, flat rocky areas with lichen and moss in coniferous and mixed woods. However, pine trees need to be present, as this is the host tree it forms its mycorrhiza with.

Herald Of Winter was first recognized as an edible mushroom toward the mid-twentieth century. Its taste is mild but still flavorful.

Due to its sliminess, it's not used as much as other mushrooms, but is still of value as one of the season's last mushrooms.

From DR. M. A. LINDBLAD'S MUSHROOM BOOK, 1920

The fruiting body often has a patch of red on it.

After some time spent at room temperature, the cap's slimy layer dries.

The flesh of the stipe is fibrous and has a satin shine.

Herald Of Winter *Hygrophorus hypothejus*

FRUITING BODY: At first covered by a slimy veil that will eventually break. The traces of the slimy ring can be seen some half an inch from where the cap attaches. A sure sign are the bright red patches that can appear here and there.

CAP: 1–2½ inches (3–6 cm) wide. To start, convex and then flat. The center is darker and often concave and the margins stay rolled in for some time. It's covered by a thick, glassy layer of slime that disappears with age. It becomes very sticky in damp weather. The color varies from a cold gray-brown to olive green and rusty brown with a fine, streaky radial pattern.

GILLS: The gills go down the stipe and are sparse and elastic. At first, a mild vanilla to butter yellow color, then turns orange.

STIPE: 1½–3 inches (4–8 cm) tall and usually thin. Light yellow to orange with a satin sheen. Slimy in damp weather and has a threadlike pattern in dry weather.

FLESH: From light to clear yellow.

SPORE PRINT: White

RANGE AND HABITAT: Occurs widespread across North America with pine trees in coniferous and mixed woods on flat, rocky, and mossy areas with lichen. Grows from October to December, though can occur into the spring in West Coast forests.

SMELL AND TASTE: Mild and when cooked, sharp and buttery.

PICKING AND CLEANING: Avoid wet weather. If the mushrooms are sticky, place them out to dry for an hour. The skin of the cap does not need to be removed.

PREPARING AND STORING: Great when fried, which gives it the best consistency. Also works well mixed with other mushrooms. Can be frozen as well as dried.

LOOK-ALIKE MUSHROOMS: The Olive Wax Cap, *Hygrophorus olivaceoalbus*, is edible and often comes earlier than the Herald of Winter. It has white flesh and white gills and the stipe has a wavy pattern. *Hygrophorus fuligineus* is a common edible mushroom in late fall in Eastern US forests under pine.

The Olive Wax Cap is often confused with the Herald of Winter.

FAIRY RING MUSHROOM

FAIRY RING MUSHROOM

THE SMALL MUSHROOMS on our front lawn have been largely ignored. They've been patiently producing new fruiting bodies—even after the lawnmower has massacred them. Until now that is. We finally realized these were the tasty Fairy Ring Mushrooms, and these days we're careful to harvest them before blitzing the lawn.

It's great to have a first class edible mushroom in your own back yard; you have total control over their health, and can see when the fruiting bodies are ready for the pan.

During some dry spells the mushrooms "wilt" and lose their shape, shrivel up, and turn pale.

However, we just need a small shower and abracadabra!—there they are again, all fresh and lithe.

Fresh Fairy Ring Mushrooms fried in butter are like candy; they taste of caramel.

(Previous page) In dry weather the Fairy Ring Mushroom is light beige.

When it is damp on the ground the mushroom is glossy and caramel colored.

Fairy Rings Packed with Spicy Mushrooms

The life cycle of a Fairy Ring Mushroom: from a small, convex cap to fully grown and cup-shaped with sparse and fully developed gills. The stipe's color changes from creamy white to cinnamon brown.

MANY PEOPLE WITH lovely yards try to eliminate their Fairy Ring Mushrooms without ever realizing what a great edible mushroom it is. Despite its small stature, it's packed with flavor. You don't need a large quantity of this mushroom to add some zing to your dish. Mushroom guru Bengt Cortin placed it as number one among all the mushroom delicacies.

THE FAIRY RING mushroom can be found nearly everywhere in the world: Europe, America, Africa, and Asia. The mushroom needs a lot of nutrients and prefers lawns and pastures in built up areas.

As opposed to most other mushrooms, this mushroom can handle several dry spells because of its tough, cartilage-like flesh. The seemingly shriveled caps are revived after a shower, which is when you want to pick them, as the stipes are chewy. The color on the cap shifts depending on the humidity and age and becomes darker.

Older specimens
turn a dark
cinnamon brown.
They are not
worth eating.

Fairy Ring
Mushrooms in a fairy
ring that might be
several decades old.

If the weather is favorable, the season will be long, from May all the way to November.

FAIRY RING MUSHROOMS grow in fairy rings or arches. It was previously believed that the formations were caused by magical powers and were created by dancing elves, witches, or other supernatural beings. They are actually created by mycelium, which grows in an ever-expanding ring, releasing new nutrients as the fungus grows in size. These fairy rings can be very old and grow between 6–19 1/2 inches (15–50 cm) a year.

Fairy Ring Mushroom *Marasmius oreades*

FRUITING BODIES: Small and slim.

CAP: At first convex, then flat to cup shaped. 1–2½ inches (2–6 cm) wide, with a low central hump that is often a darker shade. The colors vary from light beige to brown yellow. The cap margins are usually crinkled and almost transparent and streaky when the weather is damp. They turn lighter in dry weather, and regain their color and shape in more humid conditions.

GILLS: Free, sparse, and convex in the center with horizontal ridges. Creamy white at first, then dark cinnamon brown.

STIPE: Straight, 1½–3 inches (4–7 cm) tall and covered by a felt-like film that can easily be scraped off. The color is similar to the cap or slightly lighter.

FLESH: The flesh in the cap is white and elastic while the stipe's flesh is like cartilage.

SMELL AND TASTE: Pleasant. Has a similar smell to cloves, with a hint of bitter almond oil. It is milder when cooked.

SPORE PRINT: White.

RANGE AND HABITAT: Grows widespread and is common across much of North America in urban and suburban lawns and fields. Fruits from late spring to late fall and grows in rings or arches in nitrogen rich soil, fields, or pastures.

PICKING AND CLEANING: Only the cap is used as the stipe is too chewy. Divide the mushroom, as it is often host to maggots.

PREPARING AND STORING: Fairy Ring Mushrooms are spicy. When cooked fresh, they taste best fried with butter. Drying it makes the taste stronger. When crumbled it can spice up sauces and soups.

LOOK-ALIKE MUSHROOMS: False Champignon, *Clitocybe rivulosa,* and *Clitocybe dealbata,* are very poisonous. It has a similar way of growing, and will appear on lawns in arches or rings at the same time as the Fairy Ring Mushroom (page 181).

The False Champignon's cap has a frost-like belt against a dusty pink base.

GYPSY

MY WHOLE FAMILY, *including my fiancé, is wandering around the darkest, deepest woods of Hälsingland, Sweden. There are no paths, just winding trails with the smell of deer in the air.*

We hear a noise. A grouse takes off, flapping its wings heavily among the bushy branches and our fear of meeting a bear suddenly intensifies.

A large group of satiny mushrooms appear in the moss. My mother, our mushroom expert today, is sure that it's a Gypsy mushroom and a five star one at that.

The rest of us aren't so sure. They look dangerous; the cap and lamella are a brownish color, and the stipe has a ring—none of which are good signs. It's with some doubt that we place these mushrooms in our basket.

HOWEVER, OUR MUSHROOM *guide confirms that yes, it is the Gypsy.*

And so, we cleaned the mushroom, fried it with some butter, and our best bottle of wine—a Chateauneuf De Pape—was uncorked.

An unforgettable, five star experience.

◁ A group of Gypsys that have seen better days.

◻ Different ages. The ring often dries and shrivels and falls off the mature mushrooms.

The Gypsy has a bready aroma that is easy to recognize once you're familiar with it.

A Tasty *Cortinarius*

THE GYPSY WAS known as *Rozites caperata* before it was deemed to belong to the *Cortinarius* genus. *Caperatus* is latin for "wrinkled," and this mushroom certainly has a wrinkled cap.

Many people avoid *Cortinarius* because there are a few dangerous species within this genus, but there are also many edible ones. In some places, they are commonly used in food.

The Gypsy grows in large parts of the northern hemisphere and even Greenland and Japan. It doesn't seem to be bound to any specific tree type and likes coniferous, mixed, and deciduous forests.

The cap has a frosty rim and the stipe has a ring which means that beginners can be fairly certain of this mushroom when they pick it. It's an excellent edible mushroom and is worth getting to know.

The ring has dried on the older example to the left.

Here, the wavy gills can be seen.

Gypsy *Cortinarius caperatus*

FRUITING BODY: Average sized *Cortinarius*.

CAP: 2$^1/_2$–5 inches (6–12 cm) wide and fleshy. At first egg or globe shaped, then convex to flat. Beige to honey colored, usually with a silky white to violet-colored central hump. The margins are often wrinkled.

GILLS: Usually notched or widely attached to the stipe with a saw-like edge. Light beige and then cinnamon brown. Dense and wavy.

STIPE: 3–5 inches (7–12 cm) tall. The same color as the cap, only lighter. Coarsely fibrous with a wavy, scaly pattern above the ring. The base of the stipe can be swollen.

FLESH: Creamy white to beige, especially toward the base of the stipe. The cap flesh is firm and the stipe flesh is fibrous.

RING: Very visible, light ring that usually dries and falls off.

SMELL AND TASTE: Spicy and easy to recognize with a distinctive aroma. When cooked it tastes like bread.

SPORE PRINT: Rusty brown.

RANGE AND HABITAT: Found widespread across northern areas of the United States and North America, July to October, and grows in large volumes in many areas. Prefers lean, mossy, coniferous forests, preferably with spruce, but can also grow in deciduous forests.

PICKING AND CLEANING: Even the young ones are usually infested by maggots. The more mature mushrooms have a woody stipe, which can be removed if the rest of the mushroom is okay. If you're a beginner, you should check all the distinguishing features and make sure it does not have any web-like threads.

PREPARING AND STORING: Suitable to cook in most ways, both alone and with other mushroom species. Works well when dried or parboiled and frozen.

LOOK-ALIKE MUSHROOMS: Other *Cortinarius* species should be avoided, as some of them are dangerous. However, these lack a fleshy ring and instead have web-like threads—remnants of the veil (page 210).

The Deadly Webcap does not have a ring and usually has remnants from the veil on the cap margin.

SLIMY SPIKE CAP

YUCK, IT'S SLIMY! *That sure is a slick mushroom stipe and my hand instinctively pulls back.*

In a small area along a woodland path, our attention has been caught by a dozen shiny, plaster-colored mushroom caps that are spread out in the grass. Maybe they're Slippery Jacks or Weeping Boletes?

I tentatively stick my hand down again and remove a few bits of grass that are stuck on. Gingerly, I lift up the slimy fruiting body.

And so it reveals itself: the lemon yellow stipe means it must be a Slimy Spike Cap.

⬜ Younger specimens where the slimy veil has not yet detached from the cap.

◁ A mature Slimy Spike Cap, on which the typical little ledge can be seen at the top of the stipe. It was created by the veil when it detached from the cap. The dark powder is from the spores.

The fruiting bodies can be either thick or thin. This shows the progression in age from left to right. On the two younger ones the slimy layer has not yet detached and they are the least infested. The gills get darker with age.

Popular and Safe Mushroom

The rare cousin, the Rosy Spike Cap, always grows alongside the Jersey Cow Mushroom.

SLIMY! THAT'S ONE way to describe the Slimy Spike Cap without exaggerating, but it's still a popular edible mushroom. It's so easy to recognize and can't be confused with any more dangerous mushrooms.

The Slimy Spike Cap is common all over the northern hemisphere and lives in symbiosis with spruce and other conifers. Research shows that this species is more closely related to *Suillus* of the bolete order and especially to the Slippery Jack.

Apart from the Slimy Spike Cap, there are two other *Gomphidius* species in Scandinavia. One of them is the beautiful but rare Rosy Spike Cap, *Gomphidius roseus*. It's edible but should not be picked due to its rarity.

Slimy Spike Cap *Gomphidius glutinosus*

FRUITING BODY: Thick or slim.

CAP: 2–4 inches (5–10 cm) wide. Smooth and a beige pink, gray to gray violet in color. With age it develops black patches and then turns completely black. Its shape is at first convex and the margin is rolled in for a long time. It then flattens. The damper the weather the slimier it is.

SLIMY VEIL: The whole fruiting body is at first covered by a see-through, thick, slimy veil that detaches at the cap margin and rests at the top of the stipe where it creates a thicker layer. The remnants darken over time from the black spores that fall from the gills.

GILLS: Travel down the stipe and are sparse with a forked pattern. They are easy to remove. They are white at first, then gray, and finally black.

Older gills and the dried remains of the slimy ring at the top of the stipe.

STIPE: 1½–3½ inches (4–9 cm) tall. Slim to chubby. White at the top and yellow at the bottom, which tapers off. Sticky, even when cut. The surface is usually uneven and similar to coarse bark.

FLESH: White and bright yellow at the base of the stipe. Fairly soft with a fibrous stipe.

SMELL AND TASTE: Mild, even when cooked.

SPORE PRINT: Black.

RANGE AND HABITAT: Grows widespread across the cooler regions of North America together with spruce, fir, and Douglas Fir. Fruiting from August to October.

PICKING AND CLEANING: Pick when the weather is dry, otherwise it is too slimy. Use a separate basket as the mushrooms are not only sticky but will discolor other mushrooms and give them black patches. Remove the slimy layer on both the cap and the stipe, including the skin of the cap. Remove the yellow parts of the stipe, as these are woody. Avoid the mature mushrooms that have started to turn black.

PREPARING AND COOKING: Works well mixed with other mushrooms, especially spicier varieties, even when dried, frozen, or conserved.

LOOK-ALIKE MUSHROOMS: North America has several other species of *Gomphidius* that are considered edible, if not incredible. You might mistake it for a *Suillus* when glancing at it from above.

PARASOL MUSHROOM

GOING MUSHROOM-SPOTTING *on a bicycle ride usually brings rewards. Tall trees frame the winding woodland path in this mixed forest and a brisk shower has given the greenery a refreshing shake.*

We are well into July and several good edible mushrooms have started to appear, mostly chanterelles and boletes, which grow along the roadside and at the edge of the forest around this time of year.

We keep our eyes peeled and are soon rewarded.

The finest, proudest Parasol Mushroom we have ever seen appears majestically from the ditch.

We can't believe a mushroom can be this elegant!

The long, thin stipe has a wavy pattern of brown and creamy white, and the ring has not yet collapsed, which means the mushroom has recently bloomed. On the light cap, brown scales spread like chocolate shavings from the praline-shaped center.

I GINGERLY WOBBLE *the swollen foot so that it detaches from the ground. The fine stipe is surprisingly hard and strong; these qualities are essential, because it has to carry such an impressive parasol.*

◁ A recently bloomed example. The thicker part where the cap was attached forms a double-edged ring.

▯ Underneath the veil remnants, the cap's light base is covered with downy threads that radiate from the central hump and outwards.

A Parasol Mushroom That Likes the Sun

THE PARASOL MUSHROOM is a world celebrity. It appears in Europe, Asia, North America, and even North Africa. It's a gilled mushroom that many who are afraid of wild mushrooms still dare to eat. There are, however, a few similarities with some agarics, but there are also big differences between them.

The Parasol Mushroom is very tasty and slightly spicy. It's best on its own in dishes. As with most mushroom species, the younger specimens are the best, because the cap flesh is still firm and the stipe has not yet turned woody.

You can find the Parasol Mushroom growing out of plant litter in sunny and open deciduous and coniferous woods. It can even appear at the edge of forests and in grass and fields. This type of terrain is starting to disappear, which has unfortunately affected the availability of this delicious edible mushroom.

The Parasol Mushroom rarely grows alone; it's an amazing experience to see a whole host of these mushrooms in a summer field.

A Parasol Mushroom always has a light cap with dark scales and lacks a volva.

The Fly Agaric always has light veil remnants against a darker cap color and has a stipe with a volva.

Parasol Mushroom *Macrolepiota procera*

FRUITING BODY: Large and impressive.

CAP: 4–12 inches (10–30 cm) wide. Starts off egg shaped and then bell shaped until the cap margin detaches from the stipe. The base color becomes lighter with age. Dark scales appear on top when the brown surface cracks up from the smooth, central bump. The flesh is creamy white and soft.

GILLS: Detached, high up, and creamy white to beige. Delicate and turn brown with age.

STIPE: 12–16 inches (30–40 cm) tall. Long and thin. Hollow with an onion-shaped base. When mature, it has a chewy, thread-like flesh. Above the double-edged ring that often falls off, it is evenly brown. Below this, it is irregularly striped in a zigzag pattern on a light base.

RING: Very visible and has a double edge. Loosely placed on the stipe and can fall off.

SMELL AND TASTE: Pleasant and nutty, even when cooked.

SPORE PRINT: White.

RANGE AND HABITAT: Common and wide-spread across much of North America. It prefers open, sunny terrain, and the field edges of deciduous and coniferous woods. Also grows in pastures, on lawns, and on open fields, usually in groups. Fruits from July to October.

PICKING AND CLEANING: It is rarely infested. If the gills are brown, the mushroom is too old. Remove the onion-shaped base and divide the mushroom. If the stipe is chewy it can be dried.

PREPARING AND STORING: When fresh it is best fried until crispy in a hot pan. Other-wise, the mushroom can turn slimy. The caps can be fried whole and can even be breaded. If you dry it, you can make great mushroom flour from it.

LOOK-ALIKE MUSHROOMS: The Shaggy Parasol, *Chlorophyllum rachodes*, was deemed to be a good edible mushroom but is no longer recommended as such, because it can cause stomach upset. Some agarics have a similar shape and veil remnants on the cap (see page 204).

The Shaggy Parasol is distinct from the Parasol Mushroom in that the foot is smooth and lacks a wavy pattern. When damaged, the fruiting body is orange to red brown, which can be seen here where the stipe has been scraped.

SHAGGY INK CAP

EVEN AT A DISTANCE *we can see that something is wrong with our neighbor's lawn: it's been taken over by white mushrooms.*

We carefully zigzag between big and small groups of shiny, satin fruiting bodies. There is no doubt that it's the Shaggy Ink Cap that invaded this garden.

Scales covered in dew glitter like pearls in the morning sun and the caps' fur-like texture looks expensive; these mushrooms are dressed for a grand night out! A gentle wind causes the delicate elliptical and bell shaped hats to sway, and so the dance has begun.

All pumped up for a day of mushroom picking, we're slightly crestfallen to see this prey so close to home. None of us have tasted the Shaggy Ink Cap before.

Apparently the taste is similar to asparagus.

◁ Harvest them quickly! Shaggy Ink Caps can't be parboiled or dried and need to be eaten straight away.

A FEW LARGER *ones have already turned bell-shaped and the thread-like cap margin is slightly rolled up. Inky drops hang heavily from the margin, awaiting their fate.*

We escape the party, which has started to degenerate and leave slimy, black patches in the grass.

Watching the rapid deterioration of the Shaggy Ink Cap in only a few days is like watching a horror movie.

The gills are beautiful and feather-like and completely dissolve into ink once the spores mature. They are caught up by the wind before falling to the ground.

A *Coprinus* That Makes a Fleeting Visit

A perfect edible mushroom when the gills are all white. If they have started to darken you can still use the stipe.

THE SHAGGY INK cap is a frequent guest on nutrient-rich soil in parks, gardens, and roadsides. It can even be found in less pleasant spots such as garbage dumps and manure piles. The mushroom thrives off dead organic material such as soil, animal droppings, and decaying wood and has several growth periods between summer and late fall. The fruiting bodies often appear after heavy thunderstorms.

Coprinus are spread nearly all over the world. In Europe, 150 varieties are known to exist. Of these, the Shaggy Ink Cap is the best edible mushroom.

New research shows that the Shaggy Ink Cap, in addition to breaking things down, is also a meat eater; it catches and kills nematodes, or round worms. Maybe this carnivorous mushroom will becomes a natural pesticide for these parasitic worms?

Shaggy Ink Cap *Coprinus comatus*

FRUITING BODY: Long, thin, and gangly.

CAP: 1– 2 ½ inches (2–6 cm) wide. White with a light yellow/brown flower-shaped top. At first egg shaped and then long and cylindrical, finally turning bell shaped and spread out.

GILLS: To start with white, free, and dense. They then start to turn a blush color before turning black and dissolving into ink from the cap's margin and upwards.

STIPE: 2½–7 inches (6–18 cm) tall. White, thin, and hollow, with a thin ring that easily detaches.

FLESH: White and thin.

SMELL AND TASTE: Pleasant. When prepared it is considered similar to Button Mushrooms or asparagus.

SPORE PRINT: Black.

TERRAIN: Grows all over North America, usually in large quantities on disturbed soil in parks and gardens and along roadsides. It fruits most heavily in late fall and sometimes yields several harvests over the season. In some areas of the Southwest United States and more mild regions, fruiting periods occur in spring and late fall.

PICKING AND CLEANING: Only use young specimens with white gills. It is seldom infested with insects or snails. If they are pink at the edge, you can remove this part. Avoid picking in dirty or heavily trafficked spots.

PREPARING AND STORING: Prepare the mushroom as quickly as possible. Place the cleaned mushroom in lightly salted water and boil quickly before discarding the water. It is considered best stewed.

LOOK-ALIKE MUSHROOMS: None. Shaggy Ink Cap doesn't contain the poison coprine like the Common Ink Cap, *Coprinopsis atramentaria*, does. This species, which was previously deemed to be edible, should be mentioned as it often grows with the Shaggy Ink Cap; however, these days it is classified as a poisonous mushroom. The Common Ink Cap is smooth and lacks scales. It is gray to gray-black in color.

Research has shown that coprine—a poison similar to antabus—which is found in the Common Ink Cap, can cause testicular damage to rats, among other things.

BURGUNDY TRUFFLE

THE GATE OPENS into the old field in Sweden's island of Gotland. The dogs are raring to go and make a few extra leaps in the air before we let them loose and they dart off into the brushwood. They're focused on today's unusual task—finding truffles.

It doesn't take long before the eldest female marks a spot and the dog handler is right behind. Dogs love truffles as much as we humans.

Doggie treats are handed out. Good girl!

We can see the top of the truffle where the dog has scraped the surface of the soil. The mushroom is only half an inch or so beneath the surface and we carefully dig out with a weeding tool.

In a moment it's been detached and we can see that it's as big as a ping pong ball. It's a hard, bumpy, earthy lump that's seeing daylight for the first time. After a few taps and sniffs we're sure this truffle is mature.

The smell is like no other. It's an intense and magical aroma, as if someone let the genie out of the bottle.

THE GRAY WEATHER is clearing up and the mild October sun is shining down on the vegetation. The hazel shrubs shimmer a warm silver hue.

Merrily, the dogs continue their task and press their noses close to the ground. They discover truffle after truffle. Of course, one or two get eaten up before us humans catch up!

◁ The lagotto romagnolo breed comes from Italy where it's used exclusively as a truffle dog.

▢ Our clever treasure hunters find these expensive delicacies. An average size truffle weighs approximately ³/₄ –1¹/₂ ounces (20–40 grams).

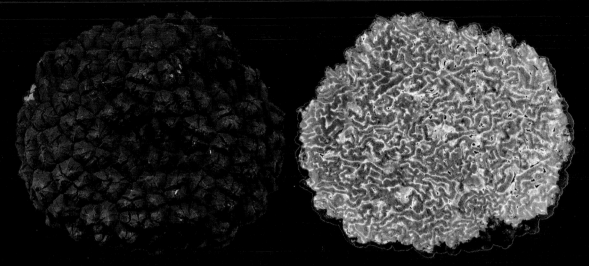

The surface looks like a pine cone, and the inside has an abstract pattern of brown and creamy white coils.

Gastronomical Gold

Farmed truffles grow on tree roots that have been implanted with truffle spores, i.e. hazel and oak trees, among others.

THE TRUFFLE HAS been beloved since time immemorial and it has a long tradition in fine dining thanks to its unique taste and smell. These days, wild truffles are more unusual, but they're farmed successfully all over the world.

This mushroom lives in symbiosis with living trees and produces fruiting bodies under the surface of the ground. The smell from the mature truffle attracts animals; the animals eat the truffle and spread the spores through their droppings.

The most sought-after edible truffles are the White Truffle, *Tuber magnatum*, and the Black Truffle, *Tuber melanosporum*, which mainly grow in southern Europe. The Burgundy Truffle, *Tuber aestivum*, is more common throughout Europe. Although it was initially thought not to grow in Scandinavia, about a decade ago it was found in Gotland and is now also farmed there. It takes about five to seven years for the fruiting body to grow. Today, the Gotland truffle is supplied to restaurants and delis, and is exported to France and America.

If you want to try picking it yourself, you can participate in a truffle safari with trained dogs and guides.

Burgundy Truffle *Tuber aestivum*

FRUITING BODY: 1–4 inches (2–10 cm) wide and can weigh up to 1 lb (0.5 kg). Brown-black in color and irregularly potato-shaped with a hard, pine cone-like shell.

FLESH: Firm. At first white, then when mature marbled in two colors with coils of creamy white to gray and cigar brown.

SPORE PRINT: Brown to golden brown.

RANGE AND HABITAT: In the ground in lime rich soil with hazel and oak. It's very rare, and in Sweden it can only be found on the islands of Gotland and Öland. It grows summer to early winter and usually in groups. Please note that in Sweden you need the land owner's permission to pick it. The Burgundy Truffle is not found in the United States.

SMELL AND TASTE: Strong and aromatic and nutty. Elevates the taste of the dish it is added to. The flavor is difficult to describe, but it's absolutely delicious.

PICKING AND CLEANING: Tastes best October through November. Remove it carefully so as not to damage the fruiting body. Brush clean with a potato brush or something similar under running water. Remove any pieces that have been attacked by maggots but leave the shell, as this can still be used.

PREPARING AND STORING: Can be used raw as a garnish, grated or thinly sliced. Add it just prior to serving as heat will ruin the aroma. A little goes a long way; for an omelet for four people, 10 grams is enough. Lasts in the fridge, wrapped first in a paper towel and then sealed in an airtight jar for eight to ten days. If it's improperly wrapped, it will flavor everything else in the fridge. Can be frozen whole or in pieces, but even then it will need to be stored in an airtight jar. Use frozen, as it will turn slimy when defrosted. Truffle butter is great and is good for adding flavor. Mix equal parts butter and truffle and add a bit of salt. Roll in parchment paper. It can be frozen in a glass jar and keeps for 3 months. The average price for the Burgundy Truffle is approximately $450 per pound. The Black Truffle costs approximately $900 per pound, and the White Truffle approximately $2400 or more.

LOOK-ALIKE MUSHROOMS: The Bagnoli Truffle, *Tuber mesentericum*, can be found on the continent, but in Scandinavia it's only found on the Swedish island of Gotland. It's not as popular as the White, Black, or Burgundy Truffles. It has a strong chemical smell that disappears after careful heating.

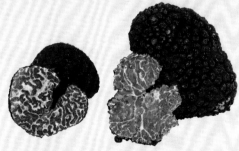

A fresh Bagnoli Truffle smells somewhat funky.

POISONOUS MUSHROOMS

THERE ARE A few old wives' tales that say all mushrooms are edible if they have a mild and pleasant taste, that animals don't eat poisonous mushrooms, and that the poison disappears if you parboil the mushroom. None of these are true. To avoid being poisoned, you have to be completely sure of your edible mushrooms and their look-alikes, and you have to get to know the most dangerous poisonous mushrooms.

The most common reason why you might have problems after eating a mushroom has nothing to do with poison; more likely, the edible mushroom was probably not handled properly. Mushrooms are very sensitive and can quickly become hotbeds of bacteria and microorganisms if they are picked without being thoroughly cleaned, or even worse, if they are placed in a plastic bag.

If you don't immediately cook the cleaned mushroom, it needs to be stored in a cool and airy place.

Mushrooms can be hard to digest and some people are more sensitive to them. When you try a new edible mushroom it's wise to start with a small portion as a test. Some people can be allergic to some of the properties in mushrooms. Also, be careful when giving cooked mushrooms to children as their digestive systems are more sensitive. There are three main groups of mushroom poison. Those that damage the cells are the most dangerous. The second group affects the nervous system and the third the digestive system.

ON THE FOLLOWING five pages we present some of the most dangerous poisonous mushrooms within these categories and those that will require medical care should you consume them. Never hesitate in seeking medical help as soon as possible if in any doubt.

- Be careful with gilled mushrooms, especially those with brown lamella.
- Avoid white mushrooms with white lamella.
- Don't pick older, infested fruiting bodies.
- Mushrooms with a volva, ring, or spidery threads can be poisonous.
- If you are picking mushrooms with others there needs to be at least one mushroom expert who can identify the mushroom.
- Be careful when buying mushrooms picked by private individuals.
- Remember that small children might put things in their mouth and there can be poisonous mushrooms in your back yard.

Call 911 if you suspect a case of poisoning. For non emergency cases and information call the American Association of Poison Control Centers (AAPCC) at 1-800-222-1222.

European Destroying Angel, *Amanita virosa*.   Death Cap, *Amanita phalloides*.

European Destroying Angel, *Amanita virosa*

FRUITING BODY: Gilled mushroom with a ring and volva.

CAP: 2–4 inches (5–10 cm) wide. As a pin head, it is surrounded by an outer veil. White to yellow in color. At first round to egg shaped, then flat, usually with veil remnants at the margins.

GILLS: White, can turn yellow.

RING: Sits high on the stipe. Easily breaks and falls off and is produced by the inner veil.

STIPE: 3–6 inches (8–15 cm) tall. Long and thread-like and has tufts above the ring. Has a swollen base with a volva that is created from the remnants of the outer veil.

FLESH: White.

SPORE PRINT: White.

RANGE AND HABITAT: This and similar, related species are found across much of North America, especially in cooler regions in deciduous and coniferous woods. Grows summer to fall.

SMELL: Faint, overbearing in the mature mushrooms.

POISON LEVEL: Deadly if not treated.

EFFECT: Damages the liver, and sometimes kidneys.

SYMPTOMS: Delayed by 8–24 hours, sometimes as long as 48 hours or even longer. Stomach pains, vomiting, bloody diarrhea, immense thirst, and so on. If not treated, can cause irreparable damage that can lead to death.

LOOK-ALIKE EDIBLE MUSHROOMS: Edible Button Mushrooms. They have faint pink to brown gills. Edible Button Mushrooms are not covered in this book.

Death Cap, *Amanita phalloides*

FRUITING BODIES: Gilled mushroom with a ring and volva.

CAP: 2–5 inches (5–12 cm) wide. As a pin-head, it is surrounded by an outer veil. At first round to egg-shaped and then flat. Varies greatly in color from gray-white to green and yellow brown and has dark radial lines from the center. Often has debris stuck to it and sometimes lots of veil remnants.

GILLS: White and dense.

RING: White, finely streaky, and thin. Created from the inner veil and often shredded or completely absent.

STIPE: 3–5½ inches (8–14 cm) tall. Long with a large and light wavy pattern. A swollen base with a volva that is produced by the remnants of the outer veil.

FLESH: White.

SPORE PRINT: White.

RANGE AND HABITAT: Though this mushroom is not native to North America, it has been introduced to but is not common along much of the West Coast and also in the Atlantic states and regions of the

Deadly Webcap, *Cortinarius rubellus*.

False Morel, *Gyromitra esculenta*.

Midwest. It should be assumed to be possible in most regions. Fruits late summer to fall in nutrient rich soil. Usually in smaller groups.

SMELL: Faint. Overbearing in the mature specimens.

POISON LEVEL, EFFECT, AND SYMPTOMS: Same as on a European Destroying Angel.

LOOK-ALIKE MUSHROOMS: Grass Green Russula, *Russula aeruginea*, doesn't have a ring or a volva and is not presented in this book. Several West Coast *Amanitas* that are edible superficially resemble the Death Cap. In addition, immigrants from Asia sometimes mistake this mushroom for edible species native to Southeast Asia.

Deadly Webcap, *Cortinarius rubellus*

FRUITING BODIES: An average sized gilled mushroom.

CAP: 1–5 inches (3–12 cm) wide. Red to yellow brown in color. Matte and felt–like to start with and then shiny, often with a hump and sometimes with velum remnants (spidery threads) on the margin.

GILLS: Wide and sparse with the same brown color as the cap.

STIPE: 2½–5 inches (6–12 cm) wide. Has a wavy pattern from the velum remnants.

Thickens towards the pointy base and sometimes has a yellow belt.

FLESH: Golden brown.

SPORE PRINT: Brown.

RANGE AND HABITAT: Fairly common all over Sweden but not so common inland in the northern parts of the country. Prefers coniferous woods and even beech forests. Grows late summer to late fall. Reported only very rarely from North America.

SMELL: Like a potato covered in soil.

POISON LEVEL: Deadly if not treated.

EFFECT: Damages the kidneys.

SYMPTOMS: Delayed, from 3–7 days and in some cases not until two weeks later. Rarely affects the stomach or intestines. At first gives a burning pain in the mouth and throat and causes an immense thirst. Then kidney pains, headache, shivers without a temperature, and so on.

LOOK-ALIKE EDIBLE MUSHROOM: None. Some cases of poisoning have occurred with careless picking of the Funnel Chanterelle and Yellow Foot.

False Morel, *Gyromitra esculenta*

FRUITING BODIES: Short and fat, like an uneven wrinkled ball.

Galerina marginata.

CAP: 1½–6 inches (4–15 cm) wide and irregularly rounded. Light to dark brown in color, usually with a rolled-in margin. Covered with wrinkly, buckled coils. The inside has irregularly shaped hollow areas.
STIPE: Irregularly flat and white to light gray.
FLESH: When cut it is tender and has a lighter color that the cap.
SPORE PRINT: Colorless.
RANGE AND HABITAT: Widespread across North America and seen generally with conifers, especially pine, though occasionally with deciduous trees. Usually in open fields and clearings that are more than two years old.
SMELL: Pleasant. If the mushroom is divided, the smell is strong and overbearing.
POISON LEVEL: Usually not as serious as the European Destroying Angel or Death Cap but can be deadly.
EFFECT: Disturbs the nervous system and can even damage blood cells and the liver.
SYMPTOMS: The reaction can be delayed, usually by 5–8 hours and sometimes up to a few days. Causes dizziness, exhaustion, double vision, and general nausea. Can also cause serious damage to blood cells and the liver. Contains mutagens that can cause long term damage.

LOOK-ALIKE EDIBLE MUSHROOMS: Black Morel *Morchella conica* (page 131).

Galerina marginata

FRUITING BODIES: Small gilled mushroom with a ring.
CAP: ½–2 inches (1–5 cm) wide. At first round, then convex to flat, often with a hump. Yellow to red brown in color. Pale when dry and cinnamon brown when wet. Has a light streaky margin.
GILLS: Dense and gray yellow to light brown in color.
RING: Thin, collar-like. Gray yellow to cinnamon colored and sometimes hardly visible or even absent.
STIPE: 1–3 inches (2–7 cm) tall. Thin, usually with white fluff at the base.
FLESH: Yellow brown.
SPORE PRINT: Brown.
RANGE AND HABITAT: Common and widespread across much of North America, in groups or bouquets. Mainly found on rotting fir or pine wood but sometimes on wood from leafy trees, in wood shavings, or along running tracks. Grows in the spring and fall.
SMELL: Like flour.
POISON LEVEL, EFFECT, AND SYMPTOMS: The same as the Death Cap and European

Royal Fly Agaric, *Amanita regalis*.

False Champignon, *Clitocybe rivulosa*.

Destroying Angel but with a lower poison level.

LOOK-ALIKE EDIBLE MUSHROOM: Sheathed Woodtuft, *Pholiota mutabilis*, which also grows on tree stumps. It has small tufts or scales that stick out on the stipe below the ring. This mushroom is not featured in this book.

Royal Fly Agaric, *Amanita regalis*

FRUITING BODY: Gilled mushroom with a ring and volva.
CAP: 3–8 inches (8–20 cm) wide. First round, then flat. Light to dark brown with creamy yellow veil remnants. (Compare with the more common lighter species on page 204).
GILLS: Dense and white to creamy yellow in color.
RING: Creamy white. Thick and looks like a bell-shaped skirt with a dark border at the bottom.
STIPE: White to creamy yellow. Has a swollen base with scaly wreaths around it.
FLESH: White.
SPORE PRINT: White.
RANGE AND HABITAT: Grows mostly in spruce woods. Does not grow in the far north or south of Sweden and grows in late summer to fall.

SMELL: Faint.
POISON LEVEL: Hospital treatment is required.
EFFECT: On the peripheral and central nervous system.
SYMPTOMS: Fast acting, after only 30–60 minutes. Stomach ache, vomiting and sometimes diarrhea. Affects the nerves with confusion, dizziness, and blurred vision among other things.
LOOK-ALIKE EDIBLE MUSHROOM: Parasol Mushroom, *Macrolepiota procera* (page 195).

False Champignon, *Clitocybe rivulosa*

FRUITING BODY: Small gilled mushroom.
CAP: 1–2 inches (2–5 cm) wide. Smooth. White or gray white to flesh colored. Flat or with a slightly concave center. Shiny in dry weather. Thin and irregularly wavy margin on the mature specimens.
GILLS: Dirty white. Attached or shortly decurrent and dense.
STIPE: Short, leathery, and thin. White to dirty white and changes to a darker shade toward the base of the stipe.
FLESH: Thin and leathery. White to dirty white in color.

Livid Agaric, *Entoloma sinuatum*.

SPORE PRINT: White.

RANGE AND HABITAT: In North America, seen only in parts of California. Prefers lawns and appears in groups, fairy rings, and arches.

SMELL: Like flour.

POISON LEVEL: Very poisonous. Requires immediate treatment with the antidote atropine, which works immediately, even with severe symptoms.

EFFECT: The peripheral nervous system is affected.

SYMPTOMS: Appear very quickly, after only a few hours, with severe headache, diarrhea, and nausea. Severe sweating and salivation, running eyes, the shakes, and more.

LOOK-ALIKE EDIBLE MUSHROOM: Fairy Ring Mushroom, *Marasmius oreades* (page 181).

Livid Agaric, *Entoloma sinuatum*

FRUITING BODY: Large, thick gilled mushroom.

CAP: $2^1/_2$–10 inches (6–25 cm) wide. Gray to creamy yellow in color and convex to flat. The margin stays turned inward for a long time. It has a central hump and is shiny with dark radial lines. Light yellow to gray yellow in color.

GILLS: Sparse and deeply notched. At first pale yellow, then pink.

STIPE: White to creamy yellow. Thick and coarse and larger at the bottom. Finely fibrous and usually with a bent base.

FLESH: White and compact.

SPORE PRINT: Pink to dirty red (page 8).

RANGE AND HABITAT: Alongside oak trees in woodlands and parks; widespread and locally common in Eastern and Central United States. Grows early summer to fall and usually grows in smaller groups.

SMELL: Strong and floury in the young mushrooms. The older ones have an oily smell.

POISON LEVEL: Very poisonous.

EFFECT: Can temporarily affect the liver and kidneys and cause serious dehydration.

SYMPTOMS: After an hour, severe stomach pains, vomiting, and ongoing, occasionally bloody, diarrhea.

LOOK-ALIKE EDIBLE MUSHROOMS: St. George's Mushroom, *Calocybe gambosa* (page 167). Edible champignons with pale pink to brown gills. Miller Mushroom, *Clitopilus prunulus*, always have a decurrent gill attachment (not covered in this book).

Acknowledgments

WITHOUT THE SWEDISH magazine *Nature & Garden* (*Natur & Trädgård*) this book would not have come about. Ever since we published our first article about the Burgundy Truffle in 2005, Sven Lindholm, the editor in chief, and Lotta Flodén, news editor, have consistently encouraged us to head out to the woods to source new materials for mushroom articles.

Over the years, chef and mushroom expert Dieter Endom has shared his valuable experiences with us and has kindly shown us his secret mushrooming spots.

Michael Krikorev, a mycologist with a vast mushroom know-how, has fact checked this book and saved us from several embarrassing pitfalls.

We share a lot of fun mushroom memories with photographer Bengt O. Pettersson, and his family. While working on this book, we have been able to tap into his expertise in photo retouching and layout.

The book has been sharpened and fine-tuned thanks to the inspiration and guidance of publisher Peter Wivall, editor Martin Ransgart, and Annika Lyth, our layout designer.

Recently I found an old mushroom book in a box at my parents' house. It's called *In the mushroom forest* (*I svampskogen*) by Nils Suber and is from the year 1950—the year I was born. The dedication to my father gave me the chills;

'To Vidar Forsberg, manager, from Nils Suber, your friend'.

Helpful Books for Mushroom Identification

FIELD GUIDES FOR THE BEGINNING MUSHROOMER

Barron, George. 1999. *Mushrooms of Northeast North America; Midwest to New England.* Edmonton, Alberta, Canada: Lone Pine Publishing.

KUO, M. 2007. *100 Edible Mushrooms.* Ann Arbor: University of Michigan Press.

LINCOFF, GARY, 2010, The Complete Mushroom Hunter: An Illustrated Guide to Finding, Harvesting, and Enjoying Wild Mushrooms. Quarry Books.

MORE COMPREHENSIVE FIELD GUIDES
ARORA, DAVID. 1986. Mushrooms Demystified: A Comprehensive Guide to the Fleshy Fungi. Berkeley, Calif.: Ten Speed Press.

BESSETTE, ALLAN E., WILLIAM C. ROODY, AND ARLENE R. BESSETTE. 2000. *North American Boletes.* Syracuse, N.Y.: Syracuse University Press.

HOLMBERG, PELLE AND HANS MARKLUND. 2013. The Pocket Guide to Wild Mushrooms. New York: Skyhorse Publishing.

LINCOFF, GARY. 1981. The Audubon Society Field Guide to North American Mushrooms. New York: Alfred Knopf.

MCNEIL, RAYMOND, 2006. Le Grand Livre Des Champignons du Quebec et de L'est du Canada. Editions Michel Quintin. This is a comprehensive book of mushrooms of Quebec, Canada.

PHILLIPS, ROGER. 2005. *Mushrooms and Other Fungi of North America.* Richmond Hill, Ontario, Canada: Firefly Books.

SCHWAB, ALEXANDER. 2013. *Mushrooming with Confidence.* New York: Skyhorse Publishing.

SCHWAB, ALEXANDER. 2007. *Mushrooming Without Fear*. New York: Skyhorse Publishing.

REGIONAL FIELD GUIDES

BESSETTE, ALAN E., ARLEEN R. BESSETTE, AND DAVID W. FISCHER. 1997. *Mushrooms of Northeastern North America*. Syracuse, N.Y.: Syracuse University Press. Covering 600 species with keys leading to photographs of the more common mushrooms.
EVENSON, VERA STUCKY. 1997. *Mushrooms of Colorado and the Southern Rocky Mountains*. Denver, Colo.: Denver Botanic Gardens.

HORN, BRUCE, RICHARD KAY, AND DEAN ABEL. 1993. *A Guide to Kansas Mushrooms*. Lawrence: University of Kansas Press. An older guide, but one that addresses midwestern mushrooms with good photos as well as many pages of additional information about mushrooms and mushrooming in the prairie states.

ROODY, WILLIAM C. 2003. *Mushrooms of West Virginia and the Central Appalachians*. Lexington: The University Press of Kentucky. Covers about 400 species in the Appalachian Mountains region. Well written and easy to use.

RUSSELL, BILL. 2006. *Field Guide to the Mushrooms of Pennsylvania and the Mid-atlantic*. University Park: Keystone Books, Penn State University Press.

TRUDELL, STEVE, AND JOE AMMIRATI. 2009. *Mushrooms of the Pacific Northwest*. Portland, Ore.: Timber Press.

INTERNET RESOURCES
www.mushroomexpert.com
www.mykoweb.com
www.rogersmushrooms.com
www.tomvolkfungi.net
mushrooms4health.com
www.mycokey.com
www.freepatentsonline.com
www.healing-mushrooms.net

OTHER PHOTOS AND IMAGES:

ENDOM, DIETER: Lethal Web Cap, pages 189 and 210.

ISTOCK PHOTO: Sponge, p. 126. Swirly Wood Cauliflower, last photo, p. 129. Swirly Black Morel, p. 132. *Morchella esculenta*, p. 133.

KRIKOREV, MICHAEL: Pine Bolete, p. 26. Livid Agaric p. 171. False Champignon p. 185. Rosy Spike Cap p. 192. Bagnoli Truffle, p. 207. *Galerina marginata*, p. 211. False Champignon, p. 212. Livid Agaric, p. 213.

SEDIN, CAMILLA: Chanterelle dog, p. 66.

WALLANDER, HÅKAN: Mycelium on a Velvet Bolete, p. 2.

Alphabetical Index

Albatrellus confluens 97, 101, 105
Albatrellus ovinus 97, 105
Amanita phalloides 209
Amanita regalis 212
Amanita virosa 209
Arched Wood Wax 173–5

Bagnoli Truffle 207
Bare-toothed Russula 151-3
Bay Bolete 15, 27, 29–31
Bearded Milk Cap 145, 149
Birch Bolete 51, 55, 57–59
Bitter Bolete 15, 23, 27
Black Trumpet 87–93
Black Truffle 206-7
Black Morel 133
Bolete Eater 14
Boletus badius, see Xerocomus badius
Boletus edulis 9–15, 23, 27
Boletus pinophilus 15, 23, 25–27
Boletus reticulatus 15, 17–23, 27
Burgundy Truffle 205–207

Calocybe gambosa 167–171, 213
Cantharellus amethysteus 66
Cantharellus cibarius 6, 61–7, 71
Cantharellus lutescens 79, 81-5
Cantharellus melanoxeros 66
Cantharellus pallens 66, 67, 69–71
Cantharellus tubaeformis 73–79,
    83, 85
Charcoal Burner 165
Chanterelle 6, 61–67, 71
Chlorophyllum rachodes 199
Chrysomphalina chrysophylla 79
Clitocybe rivulosa 185, 212
Clitopilus prunulus 213
Clustered Coral 119–121

Common Ink Cap 203
Copper Brittlegill 155–157
Coprinopsis atramentaria 203
Coprinus comatus 201–203
Cortinarius caperatus 187–189
Cortinarius rubellus 79, 210
Craterellus cinereus 93
Craterellus cornucopioides 87–93

Deadly Webcap 79, 210
Death Cap 209–210

European Destroying Angel 209
Entoloma sinuatum 171, 213

Fairy Ring Mushroom 181–185, 213
False Champignon 185, 212
False Chanterelle 67–71
False Morel 45–7, 131–133, 210–211
False Saffron Milk Cap 145, 147–149
Fenugreek Milk Cap 141

Galerina marginata 211
Giant Puffball 107–109
Gomphidius glutinosus 191–193
Gomphus clavatus 115–117
Grass Green Russula 210
Gypsy Mushroom 187–189
Gyromitra esculenta 45, 133, 210–211

Herald Of Winter 177–179
Hydnum repandum 6, 95–97, 101
Hydnum rufescens 97–101
Hygrophoropsis aurantiaca 67, 71
Hygrophorus atramentosus 175
Hygrophorus camarophyllus 173–175
Hygrophorus hypothejus 177–179
Hygrophorus olivaceoalbus 179

Hypomyces chrysospermus 14
Hypomyces luteovirens 159

Jelly Baby 79, 85
Jersey Cow Mushroom 41, 47

Lactarius deliciosus 143–145, 149
Lactarius deterrimus 145, 147–149
Lactarius helvus 141
Lactarius rufus 141
Lactarius torminosus 145, 149
Lactarius volemus 135–141
Langermannia gigantea 106–109
Larch Bolete 37, 41
Leccinum aurantiacum 51, 53–55, 59
Leccinum pseudoscabrum 59
Leccinum scabrum 51, 55, 57–59
Leccinum versipelle 49–51, 55, 59
Leotia lubrica 79, 85
Livid Agaric 2, 169, 171, 213
Lycoperdon perlatum 111–113

Macrolepiota procera 6, 195–199,
    212
Marasmius oreades 181–185, 213
Miller Mushroom 213
Morchella conica 131–133, 211
Morcella esculenta 133
Mycenastrum corium 109

Ochre Brittlegill 161
Olive Wax Cap 179
Orange Birch Bolete 49–51, 55, 59

Parasol Mushroom 6, 195–199,
    212
Pholiota mutabilis 212
Pig's Ear 121–123